Keeping the Dream Alive

Putting NASA and America Back in Space

Michael C. Simon

Earth Space Operations
San Diego, California

Library of Congress Catalog Card Number 87-81594

ISBN 0-915391-28-7

Interior design by Mike Kelly, Slawson Communications, Inc.,
San Diego, California
Cover design by Lorri Maida; Cover production by Silke Williams

Cover photo: An artist's concept of the "Dual Keel" Space
Station configuration, established in 1986 as the baseline design
for this facility. NASA hopes to have an initial version of this
Space Station operational by the mid-1990s.

All photos courtesy of NASA.

First Printing July 1987

10 9 8 7 6 5 4 3 2 1

Contents

Preface

The publication of *Keeping the Dream Alive* marks the tenth anniversary of my first job in the space business — as a student research assistant supporting NASA's 1977 summer study on Space Resources and Space Settlements. I owe that marvelous employment opportunity to the only piece of fan mail I've ever written, an enthusiastic letter I transmitted to Dr. Gerard K. O'Neill some months earlier, after reading his visionary treatise on space colonies, *The High Frontier*. Through O'Neill colleague Dr. Brian O'Leary, my letter found its way to NASA's Ames Research Center, where agency officials were managing the selection of personnel for the summer study.

O'Neill directed that 1977 study with a flourish, and with particular sensitivity to those of us on his team who were young enough to entertain thoughts of some day living on one of his "Island One" space settlements. The product of our summer's labor was a hope-filled blueprint for the large scale industrialization and habitation of the inner solar system. According to our carefully laid-out plan, 1987 was to be an important year for our enterprise. In our final report we wrote: "The lunar base begins routine launch of materials into space at the rate of 30,000 metric tons/yr shortly after sunrise at the lunar base on 1 June 1987, completing the first milestone."

By 1987, we expected, the Space Shuttle would have completed over two hundred missions, permitting an expansive manned Space Station to begin operating in low Earth orbit. With a national commitment equaling that of the Apollo era, we reasoned, 1987 could also see the first deliveries of lunar ore mined from the surface of the Moon. The destination of this material: a 2,000 ton Space Manufacturing Facility equidistant from the Earth and the Moon, which would begin operating in 1988 with an initial crew of 150 women and men.

With the memories of these early expectations in mind, the reality of 1987 is especially difficult to accept. The Space Shuttle fleet, with not two hundred but only two dozen missions to its credit, is grounded for the remainder of the decade. The large Space Station in low Earth orbit remains a distant dream; NASA's current Space Station plan is to cram four to six astronauts into a modest set of Shuttle launched tubes sometime in the late 1990s. Lunar bases and space factories? At our current rate of progress, I will not witness such developments until after I become eligible to retire — in June 2024.

The reality of the last decade is that my generation will probably never get to visit Island One. But we haven't lost sight of the dream, and that's what *Keeping the Dream Alive* is all about. The mistakes and neglect that have slowed our progress in space since the halcyon Apollo years do not have to be repeated. We can learn to operate the Shuttle properly and we can establish a Space Station program that makes some sense. Adventurers and entrepreneurs around the world are looking toward space, and NASA can still help lead us there. It is my sincerest hope that this book will offer a few sound ideas on how to make this happen.

Any acknowledgment for this effort must begin with a statement of my appreciation for Dr. O'Neill and his continuing leadership and inspiration. I must also express a debt of gratitude to the late William K. Linvill, who took a chance on me at Stanford, helping to provide me with the education I needed to pursue my career plan. Ed Bock, a participant in that 1977 study, was instrumental in enabling me to come to work at General Dynamics in 1982, and for this I am extremely grateful. Warm thanks must also be extended to Bob Bradley for his years of friendship and the many hours he spent critiquing my first draft of this book. A more recent mentor who also helped put the finishing touches on this effort is Ted Johnston. Thanks also to Mike Reeder and Mike Gass for serving on my technical review team. Finally, I'd like to express my fondest appreciation for Mike Kelly, whose knowledge of the publishing industry and dedication were vital to the successful completion of this project.

Michael C. Simon
1 June 1987
San Diego, California

Introduction

America's Space Program is in trouble. As the nation entered the second half of the 1980s, it appeared from all signs that the United States was soaring into an era of unprecedented excitement in the new frontier. The National Aeronautics and Space Administration, with fifteen Space Shuttle missions planned for 1986, was looking forward to its most productive year ever — twelve months filled with unparalleled scientific and commercial activity in space. The Shuttle would launch NASA's first two planetary exploration probes in nearly a decade: the Galileo mission to Jupiter and the Ulysses mission to the Sun. Also to be lofted into orbit by the Shuttle in 1986 was the billion-dollar Space Telescope, ten times more powerful than any telescope on Earth. During 1986, the National Commission on Space would complete an exhaustive study of the Space Program, providing a blueprint for ambitious achievements in space for the next half century. As the first step toward these grand objectives, a preliminary design for the nation's new manned Space Station was to be completed by the year's end, giving full definition to President Reagan's vision of a permanent human outpost in Earth orbit.

It was not a month into 1986, however, before fate intervened, destroying NASA's ambitious plans. In a tragedy that evoked more national emotion than any event in a generation, NASA's twenty-fifth Space Shuttle mission exploded barely a minute into flight, in view of millions of horrified spectators. The Challenger fireball took with it the lives of seven heroic astronauts and obliterated one quarter of the nation's fleet of Shuttle Orbiters. It grounded the entire Shuttle system for over two years, threw NASA's Space Station plans into turmoil, and rocked the very foundation of America's space effort. The Challenger accident turned a year of spectacles into a year of debacles. NASA's highlight for 1986 turned out to be the launch of a Navy communications satellite aboard an Atlas/Centaur rocket, a feat that would have been considered routine had the space agency's plans not fallen into total disarray.

In the months following the Shuttle tragedy, the country has struggled to get its Space Program back on track and to regain its confidence in humanity's ability to conquer the heavens. The lessons of *Challenger* will be hard-learned; recovery from this tragedy will take at least a decade and will test the resolve of mankind to seek the benefits of expanding the human presence into space. As this process of recovery unfolds, the leaders and supporters of America's Space Program must face a number of critical questions regarding the future of space exploration and the development of space resources. Most of these questions fall within the context of three general issues relating to the technology, economics, and goals of our space effort.

To resurrect the Space Program, we must decide what roles the Space Shuttle, Space Station, and other space systems should play in our conquest of space. We need to determine the most equitable and efficient way of sharing the costs of these systems between the government and private industry. And most important, we must define our ultimate objectives in space, and map out a long term strategy for using our space systems to achieve these goals. If we use these three essential questions as a guide, we can give new direction and meaning to our nation's space activities, as America renews its efforts to lead humanity into a prosperous era of large scale space development.

The Right Roles for the Space Shuttle and Space Station

The first of these three questions is the most obvious concern, in view of the doubts about our space transportation capabilities raised by the Challenger accident. There is currently widespread disagreement over how NASA's Space Shuttle and Space Station should be utilized to facilitate the expansion of mankind's activities into space. The loss of Challenger lends new urgency to this question, since it demonstrates that the Space Shuttle may be too primitive and prone to disaster to make space travel truly routine. In fact, long before the Challenger tragedy, shortcomings in the Shuttle system were becoming evident. According to John Young, NASA's most experienced astronaut, near-disastrous malfunctions occurred during at least four Shuttle missions preceding Challenger's last flight. In addition to its technical difficulties, the Shuttle is, to use the words of space visionary Arthur C. Clarke, "a financial disaster." When the Shuttle was approved by President Nixon in 1972, it was expected to fly over sixty missions per year by the early 1980s, at a cost of about ten million dollars per flight. But after ten years of development and five years of operation, the Shuttle has successfully flown no more than eight missions in any one year, and has cost the American taxpayers over $250 million every time it has been flown.

Does this mean that America's multi-billion dollar Space

Shuttle program has been a total failure? Not necessarily. The Shuttle remains one of the most technologically advanced devices ever built by man, and can still play a key role in the next decade of space development. What is apparent, however, is that NASA grossly overestimated the ability of the U.S. Government to operate a system as complex as the Shuttle. The space agency never had any chance of meeting the ambitious schedules and cost targets that were initially laid out for the Shuttle program. Even after the Challenger disaster, it took several months for top space agency officials to accept the fact that the Space Shuttle should be used only for those missions that absolutely require the vehicle's unique capabilities. It is unfortunate that the Space Shuttle failed to meet its objective of providing low cost access to space for all types of customers. But it will be even more unfortunate if the Shuttle's performance continues to be impeded by unrealistic expectations that divert the system from its most productive possible role. In its proper capacity as a vehicle for supporting research and development in space, the Shuttle can still serve as a tremendous asset to the United States and the rest of the world.

With the Space Shuttle playing a more limited and focused role than originally planned, the nation's *Space Transportation System* will have to be restructured. When this reconfiguration is complete, the Space Shuttle may represent only a small part of the overall fleet of vehicles used to provide us with access to space. An assortment of expendable launch vehicles, shunned by NASA until the recent Challenger mishap, is still available and now appears certain to have a continuing role in the national space effort. NASA's two primary customers for space launch services, the Department of Defense and the commercial satellite industry, have long expressed their desire to maintain an operational fleet of expendable rockets, which served all of America's space transportation needs prior to the Shuttle era. Now that the Shuttle has shown its fallibility, these customers are finally going to get their wish. A few months after the Challenger accident, the U.S. Congress appropriated emergency funds for the acquisition of proven expendable rockets, which will perform dozens of military space missions that had been scheduled to fly on the Space Shuttle.

While meeting near term launch needs with existing systems, we can also invest in the development of space transportation systems far more advanced than today's expendable rockets or the Space Shuttle. The basic technologies employed by expendable launch vehicles are over twenty-five years old, and the Shuttle, which was developed under severe budget constraints, incorporates few of the advances in space technology made over the past ten years. An entirely new system, such as the "Orient Express" space plane endorsed by President Reagan, might be required to augment or replace the Shuttle by the early twenty-first century. The space plane is only one of dozens of concepts for newer and more powerful space transportation systems that are currently being evaluated by NASA, the Department of Defense, and the aerospace industry. Some of the systems under consideration are manned, while others are not designed to carry human cargo. Some are reusable, some are expendable, and others, like the Shuttle, combine reused and expended elements. The Shuttle's problems have forced our nation to accelerate the development of the next generation of spacecraft, so a vehicle such as the space plane or a heavy lift cargo vehicle could be in operation as early as the 1990s.

While modernizing the nation's space transportation capabilities, NASA is likely to proceed simultaneously with the development of the Space Station, a permanently manned orbiting facility that the agency hopes to have operating by 1994. Termed the "next logical step" for America's Space Program by former NASA Administrator James Beggs, the Space Station is such a logical next step that it should have been developed twenty years ago. The Space Station will reduce the need to launch cargo into space on manned vehicles, since its crew will be available in space on a full-time basis to assemble and service equipment launched into orbit. More significantly, the Space Station will be a "stepping stone" for virtually all of our long term space activities. Its crew will conduct long duration experiments, manufacture exotic products that could not be created on Earth, and eventually operate a fleet of space based vehicles that will ferry people and equipment to the Moon, Mars, and the far reaches of the solar system. By beginning the process of moving the sphere of human activity beyond the confines of planet Earth, the Space Station will set the stage for the industrialization and colonization of space.

But the Space Station will achieve these objectives only if it is designed properly and manged effectively. The mistakes made during the development of the Shuttle cannot be repeated, or the Space Station will fall far short of its potential for stimulating the development of new and valuable uses for space. Estimates of Space Station costs and usage rates must be reasonable, rather than fabrications designed to promote the program. To make the most of the Space Station, we may even need to alter our basic philosophy regarding the role of man in space. This could generate repercussions affecting the development of Earth-based launch systems and also affect the direction of our space-based activities. Once a crew has been permanently stationed in Earth orbit, the need to expose humans to the dangers of being launched from Earth will be reduced. Equipment can be launched into space on unmanned vehicles, with the Space Station crew performing any manned functions required once this cargo reaches orbit. The need to launch people into space may arise only when it comes time to rotate the Space Station crew, or about four times each year.

In the more distant future, the Space Station will serve as a transportation terminal, with space-based vehicles routinely returning to Earth orbit with resources from the Moon and Mars. When this occurs, people living and working in space will no longer be dependent on Earth for food or other resources. NASA and the American people will have to adjust to the idea of human communities in space evolving independently of Earth, rather than responding instantaneously to the "voice of mission control." Since the use of lunar and Martian resources may become economical within the next twenty years, the Space Station must be designed to ultimately support lunar bases, colonies on Mars, and the various types of advanced transportation systems that will be needed to make routine journeys to such destinations. With the onset of self-supporting space stations, humanity will have made its last great adaptation, and the next industrial revolution will be underway.

If the Space Shuttle and Space Station are to adequately meet these challenges, the innumerable options for evolving and operating these systems must be carefully evaluated. Many issues, such as the proper role for the Space Shuttle in launching unmanned communications satellites, defy easy resolution. Another major debate within the space community concerns whether the Space Station should be manned from the outset, as opposed to starting out as an unmanned platform. Agreement has not even be reached regarding the number of stations that need to be built. The development of criteria to resolve these issues is a process to which all citizens should be invited to contribute, so the Space Program of the future serves the needs of everybody, and not just a select few individuals. If designed and utilized properly, the Space Shuttle, Space Station, and other future space systems can provide great leaps in our knowledge of the solar system and of ourselves, as well as fueling the early factories of space industrialization.

Financing America's Space Activities

The second basic question facing space planners is probably the most controversial: how will we pay for the development and operation of the space vehicles and facilities that are deemed to be desirable? The National Commission on Space recommended in its 1986 report that the United States develop six major new space transportation systems, seven orbital space stations, and permanent human settlements on the Moon and Mars by the year 2030. The commission estimated that the total cost of developing these systems over the next fifty years would be approximately three quarters of a *trillion* dollars. In absolute terms, this sum of money is enormous, but viewed in the context of some other national efforts, it can almost appear modest. The Department of Defense, for example, will spend *fifteen trillion* dollars, twenty times the amount needed to fund the space commission's program, if its budget remains at its present level over the same time span. We will undoubtedly witness a vigorous debate over whether an investment of this magnitude in space is worthwhile, but it is interesting to note that no one is questioning whether the space commission's objectives are physically attainable. The cost of space development has now replaced technological limitation as the only barrier to the large scale migration of human beings into outer space.

The high cost of space activities will probably require the U.S. Government to bankroll most major space efforts well into the next century. But does this mean that government should maintain total control over future space systems, as has been the case with the Space Shuttle? Many people within the aerospace community believe that private industry should invest more of its resources in the development and operation of space systems, and that the government should instead focus on inducing the private sector to bring the benefits of free enterprise into space. Experts on the economics of the Space Program agree almost universally that NASA's procurement process is partly to blame for the high cost of developing and operating space systems. The way most programs are managed, neither NASA's employees nor the agency's contractors are given strong incentives to reduce costs. But by making use of the many precedents that exist for government stimulation of private investment, free market forces could be given greater influence in the Space Program. If past experience is a valid indication, "privatization" could help reduce the expense of America's space systems and improve their usefulness as well.

The Space Station presents NASA with its first major opportunity to correct the mistakes that have plagued the Space Shuttle and to utilize new methods of getting the private sector involved in space development. Officials within the space agency are currently trying to reach agreement on whether the Space Station should be a wholly government-funded enterprise, or whether companies building Space Station elements should be encouraged to invest in the program. Rather than develop the Space Station through conventional aerospace contracts, NASA could utilize a variety of innovative approaches to combining public and private resources in the implementation of this project. As an example, NASA could try to induce an aerospace firm to use some of its own resources to help finance the construction of a Space Station laboratory. A company might be motivated to make such an investment if the space agency agreed to utilize the lab and reimburse the company, paying a rental fee sufficiently high to offer investment recovery. An arrangement such as this would replace taxpayer investment dollars in the lab module with private capital and would ensure operation of the laboratory by an industrial concern that had a stake in its efficient operation.

Had the Space Shuttle been developed in such a way, NASA and its contractors might have placed greater emphasis on long term economy when key decisions were being made in the concept selection process. A fully reusable spaceship would almost certainly have been more cost-effective than the partially reusable Shuttle that is in use today. With little accountability for downstream events, the NASA-industry team responsible for developing the Shuttle willingly compromised long term efficiency to get presidential and congressional approval for the partially reusable Shuttle concept, which required a more modest short term federal outlay. If the Shuttle were designed to be operated by a company whose profits depended on keeping operating costs low, a more intelligent debate might have preceded the vehicle's development. If this had happened, the companies supporting NASA's Shuttle operations today would have strong incentives to minimize the number of employees on the job, rather than the motivation to expand their contracted empires. Firms operating the Shuttle would also have greater decision-making power, enabling them to eliminate unnecessary bureaucratic expenses.

History has shown that the government can make effective use of loans, grants, market guarantees, tax breaks, and other incentives to convince the private sector to share in the financial risks and rewards of large projects. Identifying the best arrangements for financing the nation's space projects may prove to be just as important as selecting the right space vehicles and facilities. These two fundamental issues are interdependent; the availability of certain financing options may promote the development of particular types of space systems, while the availability of particular types of systems will inevitably have an impact on the rate of *space commercialization.* To make the most of the Space Shuttle and Space Station and to stimulate a true industrial revolution in space, NASA needs to get away from the way the agency has done business over the past twenty years and place greater reliance on the market forces that have fostered economic growth since the 1700's.

Establishing Long Term Goals in Space

While struggling to define the proper roles for the Space Shuttle and Space Station and to identify the best ways to fund space development, we must face a third important question: what are our long term goals in space? In their preoccupation with preparing and defending each year's space budget, many NASA planners forget that they originally joined the space agency to transform their long term dreams of space travel and exploration into reality. The vast majority of NASA employees work on space systems that are already in operation or projects that will be completed within five years. In fact, it is common within the aerospace industry for individuals interested in long range goals to be considered impractical. Consequently, there is no clear consensus within NASA as to what should be done after the Space Station is built. This is not entirely the fault of the space agency; NASA is rarely encouraged by the Office of Management and Budget or any other participants in the government budget process to define long term goals and programs.

The National Commission on Space was established by President Reagan in late 1984 largely out of recognition of the need for better definition of long range Space Program objectives. If our Space Program continues to be managed without a clear sense of long term direction, we will spend exorbitant sums of money on space vehicles and facilities that will rapidly become obsolete. A case in point is the Apollo Program of the 1960s and early 1970s. The goal of sending a man to the Moon seemed visionary when established by President Kennedy in 1961, but lack of thought by NASA regarding what would be done after we reached the Moon delayed our progress in developing space by as long as thirty years. Had the U.S. been willing to defer the first lunar landing by five or ten years, our missions to the Moon could have been accomplished by utilizing a permanent Space

Station and a fleet of reusable, space-based vehicles. Such a system could have been developed for about the same cost as the more expedient but largely dead-ended "Lunar Orbital Rendezvous" Moon landing scenario that was employed instead. As a result, we will probably have to wait until the twenty-first century before a reusable Earth-to-Moon transportation system is in place.

In response to their rare opportunities to develop visionary space concepts, NASA planners have suggested and studied a large number of long range programs. During the late 1960s, staging of a manned mission to Mars was widely considered to be the most logical encore to Project Apollo. However, this option was ruled out due to its projected high cost and questions over its benefit to the public. Under pressure to show a more tangible return on its space investments, NASA was forced to limit its major goals for the 1970s to completion of the Skylab space station project and development of the Space Shuttle. With the completion of Shuttle development and approval of a new Space Station, NASA has recently considered the long term goal of establishing a permanently manned base on the surface of the Moon. Such a base could first serve as a scientific outpost and eventually grow into an industrial facility for mining and processing lunar raw materials. Since much less energy is required to launch materials into space from the Moon than from Earth, many advocates of large scale space habitation view the lunar base as the most practical long range objective for our space program.

With the current state of space technology, either a manned Mars mission or a lunar base could be achieved within the first decade of the twenty-first century. The National Commission on Space has recommended that the United States commit itself to the accomplishment of even more ambitious goals, to provide direction for national space activities for the next half century. The programs recommended by the commission include a mixed fleet of advanced launch vehicles and Space Station elements constituting a "highway to space." Also proposed by the space commission is a spectacular "bridge between worlds" that includes an extensive space manufacturing system as well as permanent human settlements on the Moon and Mars. The basic technologies required to accomplish these goals, such as the ability to support people in space for extended periods of time, have already been developed. An even more ambitious goal, large scale space colonization, could probably be achieved by early in the next century, and would not require an annual space budget much greater than what NASA received during the Apollo era.

Whatever goals are selected to provide focus for our national space activities, they should be consistent with the fundamental objectives of our government: they should offer improvements in our quality of life, be achievable without unreasonable cost or risk, and offer flexibility in pursuing still more ambitious options for our people as our horizons expand. Without definition of such objectives, our space activities are destined to decades of drift, and the benefits of space development will in all likelihood cost more to obtain than necessary. Our ability to finance, develop, and utilize space systems in a manner that benefits all of humanity will depend largely on our success in defining these long range goals.

Organization of this Book

The intent of this book is to help its readers develop a greater appreciation for what we are capable of achieving in space and what is required to get us there. The ten following chapters are divided into three sections, each of which addresses one of the three subject areas summarized in this introductory chapter. Section I, *Gateway to Space,* examines the history of the Space

Shuttle and the early phases of the Space Station program. Section II, *Paying the Toll,* discusses the impact of economics on the Space Program and presents ideas for reducing the cost of space development. The third and final section, *Beyond the Gateway,* addresses the role of humans in space and describes some of the ambitious space goals that could be accomplished within the next fifty years.

The reader should note that all cost figures presented in this book are in constant (inflation-adjusted) 1987 dollars, unless otherwise indicated.

I

Gateway to Space

America's space program is frequently compared with this country's nineteenth century drive to explore and populate the western frontier. Daniel J. Boorstin begins his Pulitzer Prize-winning history, *The Americans: The Democratic Experience,* with a description of the last century's "go-getters," who ventured westward and "made something out of nothing...brought meat out of the desert, found oil in the rocks, and brought light to millions." These go-getters were motivated by restlessness, curiosity, entrepreneurial spirit, and the desire to exercise their freedom to the fullest possible extent. Faced with the same passions, the "go-getters" of the late twentieth century are turning their sights upward, in the hopes of visiting unexplored planets, obtaining riches from the Moon and asteroids, beaming megawatts of solar energy down to Earth, and setting up frontier colonies throughout the solar system.

The conquest of space, however, will be far more costly than the taming of the American frontier. Today's spacefaring adventurers cannot simply hitch a mule and ride out to the heavens. Placing even the smallest quantity of material in Earth orbit requires the harnessing of incredible amounts of energy and costs more than nearly any individual can afford. Cognizant of this problem, the National Aeronautics and Space Administration has attached its highest priority to increasing the accessibility of space. The space agency is now in the midst of a thirty year effort to develop a system to reduce the costs of doing business in space to more reasonable levels. The two main elements of this system are the Space Shuttle, which NASA began operating in 1981, and the Space Station, which the space agency expects to launch and assemble in orbit during the mid-1990s.

Just as the development of railroads precipitated a dramatic increase in western expansion during the last century, the architects of the U.S. Space Program hope that the Space Shuttle and Space Station will represent the first elements of an expanding, operational system, enabling America's first "space go-getters" to live and work in space. If the railroads proved to be America's gateway to the western frontier, then the Space Shuttle-Space Station system can be viewed as our "Gateway to Space," designed to provide private citizens and companies with access to space for at least the remainder of the twentieth century.

The idea of developing a station in Earth orbit and a reusable spacecraft to service it is not new. In fact, such a system was originally proposed years before NASA even existed, by the space pioneers whose vision gave birth to the U.S. Space Program. But while the shuttle concept was conceived in the 1950s, America was sidetracked from its efforts to develop a reusable spacecraft by the great "Moon race" of the 1960s. The U.S. resumed its progress toward this goal by initiating the Space Shuttle program in the early 1970s. The Shuttle began operating in the spring of 1981 and will remain this country's sole means of putting people into space until at least the year 2000. The other element of our Gateway to Space, the Space Station, has an even grander heritage. The rotating wheel space station concept popularized in the early 1950s by Wernher von Braun was actually first proposed by Hermann Noordung in 1929. Still earlier, the Russian scientist Konstantin Tsiolkovsky described a manned space station that would harness the power of solar energy, rotate to provide artificial gravity, and utilize space greenhouses to grow food for its residents. The year: 1903.

With the completion of the Shuttle development program and NASA's recent selection of the companies that will design and build the Space Station, we have now progressed more than halfway toward the realization of a Gateway to Space. The tragic *Challenger* accident of early 1986, however, has called into question both our goals in space and the means we have chosen to achieve these goals. We now realize that the Space Shuttle will probably never be operated as routinely as a railroad, or even a commercial airliner. Designed to fly once a week at a cost of about ten million dollars per flight, NASA would now gladly settle for a Shuttle that could be flown once a month for a cost of one hundred million dollars per mission.

NASA's failure to achieve the performance and cost goals of the Space Shuttle program has demonstrated that the space agency's original expectations for the Shuttle program were grossly optimistic. It also exemplifies the fact that we are still in a very early stage of space travel, and that past comparisons of the Shuttle with commercial airliners were highly inappropriate. A few hours after the Challenger tragedy, California Space Institute physicist David Brin observed, "We are just past the daredevil phase (of space transportation development) but have not yet reached the stage of the DC-3. We are still taking risks to demonstrate the usefulness of space, which will become apparent to everyone within the next 15 years when we have the space equivalent of the DC-3." Despite its shortcomings, the Shuttle has succeeded in demonstrating the usefulness of space, performing twenty-four stunning and successful missions prior to the Challenger accident. These triumphs have given us hope that the system can achieve its most important long range goal: to support the establishment of mankind's first permanent foothold in space.

The success of the Space Shuttle is a miracle, considering the budgetary and schedule constraints imposed upon NASA by a succession of presidents, Congress, the Department of Defense, and the Shuttle's commercial customers. Had any of these parties been willing to pay their fair share of the cost of achieving routine access to space, NASA might have had the people and equipment necessary to build the optimum Space Transportation System. For our Gateway to Space to be achieved under existing conditions, NASA must resume safe operation of the Space Shuttle, which is the most obvious and urgent task facing the space agency today. Less obvious but equally important is the need for us to understand why the Shuttle's problems occurred and to ensure that past mistakes are not needlessly repeated as we move on to future programs. By achieving these two important goals, we can move ahead with the Space Station program, assured of the best possible prospects for the successful establishment of our Gateway to Space. Through this gateway we can then embark upon the progressively more ambitious programs that will some day open up the frontier of space to all of humanity.

1

The Era of the Space Shuttle

America's Space Program emerged from infancy on April 12, 1981, as astronauts Robert Crippen and John Young climbed aboard the Orbiter *Columbia* for the first-ever ride on a reusable spaceship. The era of the Space Shuttle was about to begin, twenty years to the day after Soviet cosmonaut Yuri Gagarin had become the first human space traveller. It was a fitting anniversary to commemorate, for the Soviet achievement two decades earlier had set in motion a series of events that would shape the United States manned space program for years. Within days of Gagarin's historic flight, President John F. Kennedy had issued an urgent memorandum, directing Vice President Lyndon B. Johnson to identify a space program that would enable the U.S. to "beat the Soviets" and provide "dramatic results in which we could win." Six weeks later, the President made his now-famous proclamation committing the United States to accomplish a manned mission to the Moon by the end of the 1960s.

By taking a stunning early lead in the development of space technology, the Soviet Union motivated the United States to devote its first ten years of manned space activities to the achievement of dramatic victories. Project Apollo fulfilled America's desire to demonstrate superiority in the new frontier, but after achieving this victory the national euphoria over winning the "Moon race" wore off rapidly. NASA was forced to curtail its Apollo plans, and instead devoted its agenda for the 1970s to showing that there were more practical reasons to send people into space. To lead the nation into this new era of space *exploitation,* NASA elected to develop the Space Shuttle. The Shuttle was designed to be the embodiment of practicality: it would be versatile, inexpensive to use, and durable enough to serve as the workhorse of a new industrial age in space.

A Spaceship is Born

The idea of developing a reusable space transportation system was originally conceived as a means of supporting the classical "rotating wheel" space station concepts proposed in the 1950s by early space pioneers such as Wernher von Braun. In its efforts to define a practical space project to succeed Project Apollo, NASA proposed in 1969 to achieve low cost access to space by using reusable chemical and nuclear rocket transportation systems. To help build its case for development of such a vehicle, NASA sought the support of the Department of Defense, whose satellites were expected to represent at least one third of the nation's launch needs during the 1980s and beyond. In 1971, NASA obtained the Defense Department's blessing by agreeing to develop a vehicle of sufficient size and capability to launch large military satellites. On January 5 of the following year, NASA Administrator James Fletcher briefed President Richard Nixon on the space agency's proposed new transportation system. At the conclusion of Fletcher's briefing, the White House announced its official approval of the Space Shuttle program.

From the start, economics played a key role in the evolution of the Shuttle program. The whole idea behind the Space Shuttle was to reduce space transportation costs and risks, thereby giving the government and private industry routine access to space. In 1970, NASA hired a consulting firm, Mathematica, to analyze the cost-effectiveness of several different shuttle concepts. While a fully reusable space vehicle would almost certainly have been more economical in the long run, NASA was under pressure by the Office of Management and Budget to select a vehicle with lower up-front development costs. The current "thrust-assisted orbiter shuttle" concept was found by Mathematica to be the "economically preferred concept;" it would cost about half as much to develop as a fully reusable shuttle and would still be economical to operate. According to NASA's early cost estimates, the partially reusable vehicle could be flown for less than ten million dollars per mission, or about $150 per pound of payload placed in Earth orbit (in 1971 dollars).

This cost was approximately one tenth of what it cost to launch payloads into space with the expendable rockets in operation at the time. By achieving these economies, it was anticipated that the Shuttle would open up a new age of commercial space development, encouraging private companies to expand their uses of space for provision of such services as satellite telecommunications, weather forecasting, and the production of unique materials in zero gravity. By flying a laboratory module in the Shuttle's large cargo bay, the vehicle could also be used for low cost space experimentation. By emphasizing the Shuttle's versatility, NASA hoped to sustain widespread public interest in the program, a requirement for maintaining the support of the officials responsible for keeping the project funded.

Government approval of the Space Shuttle project, however, did not forestall the decline in overall Space Program funding that was already well underway. During the peak years of Apollo program funding, 1964 and 1965, the space agency had received a share of the federal budget equivalent to about twenty billion of today's dollars. In 1966, however, NASA's budget began a decade of steady decline. By 1975, annual appropriations for civilian space programs had been reduced to about $7 billion (in inflation-adjusted dollars), and have remained at about that level ever since. Over the span of ten years, NASA's share of government spending was reduced from nearly three percent of the federal budget to less than one percent. It is ironic that the government's purse strings were readily opened to finance grand adventures to the Moon, but were pulled tight as NASA attempted to develop a space program that would yield greater material benefits.

Even though NASA's budget was cut sharply as Project Apollo was phased out, the Space Shuttle development program got off to a good start and proceeded smoothly for several years. The early 1970s was in fact a time of considerable success for the nation's Space Program, with the completion of three highly successful Skylab missions, the Apollo-Soyuz Test Project, and the early phases of the Space Shuttle design effort. Responding to a massive letter-writing campaign by fans of the television series *Star Trek*, NASA gave the name *Enterprise* to its first Shuttle Orbiter test vehicle, which the agency and Shuttle prime contractor Rockwell International proudly unveiled in 1976. While they were not as dramatic as the exciting manned lunar missions of the late 1960s and early 1970s, these accomplishments were significant and would eventually have a great long term impact on the national space effort.

After achieving its victories of the early 1970s, however, NASA entered one of its most difficult periods. The Shuttle flight test program, which had begun very smoothly, began to run into development problems in 1977. At least a half dozen major failures were encountered during tests of the Orbiter's main engines from September 1977 through the end of 1979. Highly publicized difficulties also arose during development of the Orbiter's heat-resistant thermal protection tiles. Problems such as these are inevitable when technologies of such advancement and complexity are being developed, but they are costly and embarrassing nonetheless. The Shuttle's difficulties resulted in significant cost overruns and forced a two-year delay in the program.

While the nation waited impatiently for the inaugural flight of its new spaceship, a six-year lapse in manned spaceflight left NASA with little positive publicity to offset the mounting criticism of the Shuttle program. As if to symbolize the woeful state of the Space Program, Skylab came tumbling down to Earth on July 11, 1979 — just nine days before NASA was to celebrate the tenth anniversary of the Apollo 11 lunar landing. One of the Space Shuttle's earliest missions was to have been a flight to rescue Skylab, which NASA had hoped to use as the first building block for a permanently manned Space Station. The loss of Skylab was a humiliating end to what had been NASA's most successful program of the 1970s, and heightened America's frustration over the delays in Space Shuttle development.

Supporters of the Space Program, however, did not lose hope, and waited eagerly for the arrival of the Space Shuttle to vindicate their enthusiasm. As the target date for NASA's first Shuttle flight drew nearer, signs of increasing public interest in space began to emerge. Numerous space interest groups were formed across the country, representing one of the largest grassroots activities in America since the Vietnam War protest movement. The success of space-oriented movies like the *Star Trek* and *Star Wars* series seemed to offer an additional indication that the nation's interest in space was heading toward a period of revival. While NASA generally maintained a low public profile during this period, plans for advanced projects such as the Space Station and a return to the Moon were quietly and confidently developed, with the Space Shuttle accepted routinely as an expected fact of life for the 1980s.

The Design of the Space Shuttle

Through its Space Shuttle development program, NASA aspired to meet the needs of virtually every kind of spacefaring customer. NASA also sought to satisfy these various customer requirements economically, and to develop a system that would achieve these objectives within a tightly constrained budget. By attempting to meet all of these potentially conflicting goals, the space agency rejected the idea of developing a Shuttle optimally designed for any one particular mission. Through a process of careful compromise and selective adaptation of advanced technologies, NASA hoped to design a Shuttle that would offer distinct advantages to anyone wanting to perform work in space. There was little doubt that when completed, the Shuttle would represent the greatest engineering feat of all time. The Shuttle would combine the enormous power required to lift over a hundred tons into Earth orbit, the ruggedness needed to keep a human crew alive in space for over a week, and the delicate grace of a glider for its unpowered return to the surface of our planet. All this would have to be achieved time and time again with near-perfect reliability, for everyone knew that in the space launch business, the results of a single failure could be catastrophic.

Long before Columbia's maiden voyage, the basic elements of the Space Transportation System, NASA's official name for the Space Shuttle, had become familiar to many Americans.

The heart of the system is the Orbiter, whose construction is supervised by the Shuttle's prime contractor, Rockwell International. Final assembly of the Orbiter is performed at Rockwell's plant in Palmdale, California, but parts of the Orbiter are built by subcontractors at locations all across the country. The Orbiter's wings, for example, are built in Long Island, New York, by Grumman Aerospace Corporation. Also built in New York are the IBM computers that control the Orbiter's vital functions. The rest of the Shuttle's California connection includes Lockheed Missiles and Space Corporation, which builds the heat-resistant tiles that prevent the Orbiter from burning up during reentry into Earth's atmosphere, General Dynamics' Space Systems Division, which builds the Orbiter's mid-fuselage, and Rockwell's own Rocketdyne Division, which developed and builds the Orbiter's main engines. A number of important Shuttle accessories are even built in locations outside of the United States. These include the Remote Manipulator System (RMS), the robotic arm used to deploy and retrieve Shuttle payloads, which is built by SPAR Aerospace, Ltd. in Canada. Another critical Shuttle element built outside the United States is the Spacelab module, which was built by the nine member countries of the European Space Agency.

The largest element of the Shuttle system is the 202 foot-long, forty ton aluminum external fuel tank (ET), which contains the one and a half million pounds of propellants consumed by the Orbiter's main engines during launch. The external tank, which is built in Michoud, Louisiana by Martin Marietta Corporation, was designed to be expended after each launch. This relatively simple design approach was selected to reduce Shuttle development costs. It was determined in 1971 that the alternative, development of a reusable "flyback booster," would exceed the Shuttle funding limits imposed by the Office of Management and Budget by about five billion dollars. Attached to the external fuel tank during the first two minutes of launch are the twin solid rocket boosters

(SRB's) that provide the tremendous thrust needed for initial lift-off. The SRB's are built in Utah by Morton Thiokol, Inc.

To achieve the great velocities needed to get into orbit and to execute the delicate maneuvers required to support missions in space, the Shuttle was designed to employ four completely independent propulsion systems. In initiating the chain of events needed for a successful launch, the Orbiter's three main engines are fired for six seconds before the Orbiter leaves the launch pad. Each engine is nearly fifteen feet in height and provides over 400,000 pounds of thrust. The much-publicized problems encountered during the development of the main engines were due to the fact that the engines are probably the most technologically advanced element of the Space Shuttle system. To achieve the Orbiter's payload lift requirements, the main engines had to be designed for unprecedented size and performance. To achieve this performance, the engines must be fired at extremely high temperatures and great chamber pressures, generating problems that are compounded by the fact that the engines have to be reused for several missions.

The main engines are fired for several seconds prior to Shuttle lift-off to provide on-board computers with an opportunity to detect potential problems in the operation of the engines. The command is given to activate the solid rocket boosters only if the computers determine that the engines are working properly. The boosters are filled with a highly volatile rubber-based compound that releases an enormous amount of energy when burned. This energy provides the SRB's with the five million pounds of thrust needed to propel the Space Shuttle through the relatively dense lower layers of the Earth's atmosphere. Within approximately two minutes, the Orbiter achieves an altitude of approximately twenty-four miles and the task of the SRB's is completed. The spent solid motors are then separated from the external tank by small explosive charges. In a design compromise aimed at reducing Shuttle operations costs, the SRB's were designed to be reused for at least twenty missions. After SRB separation, parachutes are activated to cushion the fall of the SRB's to the ocean's surface, where the rocket casings are subsequently retrieved by a NASA ship, towed to shore, and refurbished.

After the SRB's are separated, it is up to the Orbiter main engines to finish the job of accelerating the Space Shuttle to orbital velocity. The Orbiter achieves this speed, over 17,000 miles per hour, after only eight minutes of flight. This critical juncture in the Shuttle launch phase is known in NASA parlance as "MECO," which stands for main engine cut-off. If any of the main engines fail prior to MECO, the Shuttle might not attain its desired orbit, forcing its astronauts to attempt a highly dangerous return of the Orbiter to a contingency landing site on Earth. When MECO is achieved, the explosive bolts connecting the huge external tank to the underside of the Orbiter are detonated, sending the ET on a return path of fiery disintegration toward hopefully uninhabited regions of the Earth's surface. Now and for the rest of its mission, the Orbiter is on its own.

The third mode of propulsion to be used during the Shuttle mission is provided by the two Orbital Maneuvering System (OMS) engines, which are located at the tail section of the Orbiter above the main engines. The OMS engines, which are powered by a highly reactive and toxic fuel known as hydrazine, are used to raise the Orbiter from its initial "parking" orbit to its ultimate orbital altitude, which can range from 130 to over 300 nautical miles. For most Shuttle missions, an altitude of about 160 miles is sufficient, but certain missions, such as those employing astronomical instruments that are sensitive to atmospheric interference, require higher orbits. Since the Shuttle's capability to lift objects into orbit is greatly reduced at higher altitudes, the orbit selected is usually as low as possible, given the constraints of the particular mission.

On numerous occasions during each mission, the Orbiter is required to use its fourth Shuttle propulsion system: a set of small attitude control thrusters, also fueled by hydrazine, that are used for fine tuning of the Shuttle's orbital trajectory. Responding to commands from the on-board computers or manual controls, these thrusters keep the Orbiter oriented properly, and are used to provide small velocity changes when the Orbiter needs to approach or depart from the vicinity of a free-flying payload or satellite. The attitude control thrusters also work in conjunction with the OMS engines to perform the Shuttle's critical deorbit maneuvers, which place the Orbiter in the proper position for its return to Earth at the conclusion of each mission.

The Space Shuttle's propulsion systems were designed with little margin for error; they must all work flawlessly to get the Orbiter through its two most dangerous phases: launch and reentry. But for the Shuttle to serve effectively as a laboratory, factory, and workbench in space as well as a transportation system, many other critical systems must also function properly. The Orbiter was designed to support a crew of at least seven people in space for a week at a time, and must provide for the complete physiological needs of these astronauts. To ensure the astronauts' safety, the Shuttle's crew compartment had to be designed to offer complete protection against the tremendous forces of launch, the hostile environment of space, and the heat of reentry into Earth's atmosphere. Within the modest living space afforded by the split-level crew quarters (flight deck above and middeck below), the Orbiter must also provide many of the amenities of home, including a shower, sleeping bags, and a galley offering dozens of space age entrees.

Other items crowded into the Orbiter crew compartment are the tools and devices needed to perform the many on-orbit tasks and experiments that require hands-on attention by the astronauts. The Orbiter's modest living space also serves as a miniature space laboratory and factory. Many life science experiments are performed in the same areas in which the astronauts eat and sleep. NASA even crams experiments into the middeck lockers that are used to store the astronauts' clothing and other personal items. In fact, some of the most important Shuttle experiments are designed to be performed in the middeck cabin, such as a series of electrophoresis demonstrations aimed at leading to the commercial production of pharmaceuticals in space.

While the Orbiter's crew compartment is alive with activity from launch through mission completion, the focus of each mission is usually what takes place in the Shuttle's mammoth cargo bay. The cargo bay's sixty foot length was originally sized for large military satellites and its fifteen foot diameter was selected to support manned activities in pressurized modules. As its contribution to the Shuttle program, the European Space Agency underwrote the billion dollar cost of developing the Spacelab, a habitable research laboratory designed to be taken into space in the Orbiter cargo bay. A docking tunnel is used to connect the Spacelab with the Orbiter crew compartment, providing Shuttle mission specialists with easy access to the lab. By enabling experiments to be performed around the clock in a shirtsleeve environment, it was expected that the Spacelab would satisfy manned space science requirements until a permanently manned Space Station could be developed. Spacelab modules of two different sizes have been developed; the "long module," at 37 feet in length, is nearly as large as the science modules that will be brought to orbit in the 1990s by the Shuttle and attached to the Space Station.

The major shortcoming of Spacelab is that it can accommodate experiments for a duration only as great as the length of a Space Shuttle mission, about one week. During the Shuttle's development, NASA considered developing a "power extension package" to extend the Orbiter's stay-time in orbit to one month, but NASA was unable to obtain funding for this project. NASA life science experiments or other investigations requiring human interaction will therefore be limited in duration to one week until the Space Station is available. For experiments that can operate automatically in an unpressurized environment, an interim solution is available: the Long Duration Exposure Facility (LDEF). The LDEF, a structure similar in size and shape to the long Spacelab module, was designed to house dozens of small automated experiments. It can be deployed from the Shuttle cargo bay with the Remote Manipulator System, and float freely in space until picked up and returned to Earth several months or years later by a future Space Shuttle mission. The LDEF is particularly suitable for very delicate experiments that can be ruined by the vibrations caused by the activity on board the Orbiter.

The Shuttle's Remote Manipulator System was developed to provide NASA with the ability to retrieve spacecraft such as the LDEF from orbit for servicing, repair, or return to Earth. This capability was intended to be used primarily for the servicing of expensive scientific and defense spacecraft, many of which orbit the Earth at altitudes accessible to the Shuttle. The RMS can also benefit commercial Shuttle customers, whose serviceable payloads include communications satellites, remote sensing spacecraft, and platforms for processing of materials in zero gravity. To further support commercial Shuttle customers, NASA developed a variety of "upper stage" rocket boosters. These expendable boosters, delivered to space in the Shuttle cargo bay, are designed to boost payloads, primarily communications satellites, to orbits that are beyond the range of the Shuttle.

Despite the drama of its launch and the versatility of its orbital operations, the most impressive aspect of the Space Shuttle's design may be the manner in which the vehicle returns to Earth. After being nudged from orbit by its OMS engines, the Orbiter dips into Earth's atmosphere, where the 2,000 degree heat of reentry quickly turns many of the Shuttle's 20,000 protective tiles red hot. Each of these tiles had to be individually designed to fit to the contour of the Orbiter. The tiles also are made of extremely lightweight materials so that the weight of this thermal protection system does not significantly reduce the Shuttle's cargo-carrying capability. From experience gained during its early manned space missions, NASA knew that the heat and ionization generated in the vicinity of the Orbiter during its plunge through the upper atmosphere would completely block off communications with the spacecraft, requiring ground control personnel to take it for granted during these few agonizing moments that the Shuttle tiles are performing their job.

Once the Orbiter's velocity has been slowed sufficiently by the Earth's atmosphere, the heat rapidly dissipates from the protective tiles and the one hundred ton Orbiter becomes the world's heaviest and most complex glider. Without the use of any kind of propulsion after its reentry, the Shuttle must decelerate from over twenty times the speed of sound to a landing speed of approximately two hundred miles per hour within a matter of minutes. The Orbiter must also land on target during its first and only attempt, since it cannot return to the skies and a second chance. To accomplish its unpowered landing, the Orbiter performs a number of graceful turning maneuvers on its glide to the Earth's surface. If the Space Shuttle were properly designed, it would be demonstrated first with the roar of its launch, and ultimately with a soft and silent return to its point of origin.

The Dawn of an Era

When the date of the Space Shuttle's maiden voyage finally arrived, the U.S. Space Program clearly had a lot riding on the first launch of its new system. With mounting skepticism from NASA's critics and heightened expectations from its supporters, a successful mission seemed critical to the space agency's very survival. While the Shuttle was expected to eventually provide more or less "routine" access to space, the drama of Columbia's first launch rivaled that of America's space shots of the 1960s. Columbia's two pilots, Robert Crippen and John Young, would be flying two billion dollars worth of hardware that represented the future of the entire American space effort. To add to the suspense, not only would they be the first Americans in space in over half a decade — they had earned the privilege of being the first two humans to ride a rocket that had never before been tested in flight.

To further heighten the suspense, Columbia had endured one final delay, while on the launch pad two days earlier. With only thirty-one seconds to go before launch, a computer malfunction halted the countdown, extending the wait of the hundreds of thousands of people who had camped around Cocoa Beach, Florida to watch the drama unfold. Finally, shortly after dawn on April 12, 1981, the Columbia's three main engines ignited, drenching Launch Complex 39A with fire and returning the roar of manned spaceflight to Cape Canaveral. The Orbiter's five computers, which were working perfectly this time, determined that the engines had reached full thrust, thereby permitting the command to be sent to ignite the Shuttle's two solid rocket boosters. Within a split-second, the commotion beneath the Orbiter main engines was completely dwarfed by a massive plume of fire and smoke, and Columbia, boosted by the six million pounds of thrust of its full propulsive complement, soared skyward to the cheers of the throngs below.

Within eight minutes, the Columbia was in orbit, over one hundred miles above the Earth. The Orbiter's cargo bay doors were gently opened to release the heat generated by the spacecraft's onboard power systems. But just as Shuttle success celebrations were getting underway on Earth, the Orbiter's on board television cameras zoomed in on the aft end of the Orbiter and revealed a potentially major problem: several of Columbia's protective tiles, needed to shield the Orbiter and its crew from the intense heat of reentry, were missing, presumably jarred loose during launch.

Ground crews quickly went to work to analyze the problem. Secret military cameras were used on Earth to photograph the underside of Columbia; the black-painted tiles at this location were particularly crucial. After several hours it was decided that the Columbia's return to Earth, scheduled for two days later, was not in jeopardy, and the mission continued as planned. The only other noteworthy problem encountered during flight was some difficulty the astronauts encountered when they attempted to close the Orbiter's cargo bay doors. The extreme changes in temperature to which the Orbiter had been subjected in space had warped the doors, preventing them from closing securely. If the doors could not be closed properly, the Orbiter would not have been able to return to Earth safely. Fortunately, the Columbia's astronauts were able to solve this problem cleverly by rotating the Orbiter slowly, "like a hot dog," as one of the astronauts put it, to normalize the temperatures.

As Columbia's first mission progressed and its satisfied witnesses filtered away from the Cape, yet another immense crowd was gathering at the other end of the continent, to watch Columbia land. Thousands of spectators crammed the roads in the California desert near the Shuttle landing strip at Edwards Air Force Base, not far from where Chuck Yeager had broken the sound barrier in his X-1 rocket plane a third of a century earlier. Only minutes after reentering Earth's atmosphere at twenty times the speed of sound, Columbia appeared over the California coast with a sonic boom, and glided to a picture-perfect landing on the now-famous dry lake bed at Edwards, to the delight of the multitudes who had come to watch.

Columbia's landing brought to a successful conclusion the first of four vitally important test flights for America's new reusable spaceship. The Shuttle's four test missions, designated STS-1 through STS-4 by NASA's new Office of Space Transportation Operations, seemed to prove that the Shuttle was a flightworthy, workable system, and provided initial encouragement that the Columbia and her sister ships — Challenger, Discovery, and Atlantis — would be able to meet the nation's space transportation needs for the following two decades. Some nagging problems were encountered during these first historic flights, but NASA engineers were adept at solving them. It was determined, for example, that shock-wave damage to the launch pad caused by the Shuttle's exhaust at lift-off could be reduced by spraying thousands of gallons of water onto the pad during launch. To solve the problem of missing Shuttle tiles, a better glue was found. According to popular legend, the miracle substance was denture adhesive.

As the Shuttle's early difficulties were resolved, a number of dramatic feats were accomplished that appeared to vindicate NASA's investment in the nation's new resource. The Shuttle's ability to retrieve payloads from orbit was demonstrated on one of its earliest flights, the well known "Solar Max Repair" mission. The Solar Max satellite had been launched by a Delta rocket in 1978, with the objective of observing the sun for a period of several years. Shortly after launch, however, the satellite's power system had malfunctioned, leaving the spacecraft tumbling uselessly in space. In April 1984, after deploying the first LDEF module, the crew of the eleventh Space Shuttle mission performed the long awaited rendezvous with Solar Max, 250 miles above the Earth's surface. The Shuttle astronauts returned Solar Max to the Orbiter cargo bay and performed two lengthy repair operations on the craft. After it was confirmed that the satellite had been successfully repaired, the RMS was used to return Solar Max to its proper orbit.

Early Space Shuttle flights such as the Solar Max Repair mission redefined man's role in space, a definition that would evolve with each new milestone the Space Transportation System would achieve. Man would no longer be just a passing visitor to space, he was now an active worker in the new frontier. In the not-too-distant future, he would surely become a permanent resident. Of more immediate importance, the early successes of the Space Shuttle helped restore confidence in the Space Program, and in so doing provided a kind of excitement that the American public had been missing for several years. A generation weaned on *Star Trek* and *Star Wars* had waited patiently, and were given a youthful glimpse of their space age dreams becoming reality.

2

The Selling of the Shuttle

The early successes of the Space Shuttle gave NASA a much needed boost, helping the space agency to earn the confidence of potential Space Transportation System customers. From the start, NASA realized that the Shuttle program would be successful only if the Department of Defense and private industry remained loyal users of the nation's new launch system. Without maintaining a broad base of customer support, NASA would find it extremely difficult to sustain the political support needed to continue the Space Shuttle's multibillion dollar funding. But by the time the Shuttle became operational, the space agency was faced with an even more pressing need to find as many customers as possible. The number one goal of the Shuttle program was to reduce the cost of space transportation, and only at very high flight rates would the STS have any chance of meeting this objective. Obsessed with the desire to fly as many Shuttle missions as possible, NASA stressed the system to its limits and ventured down a path toward inevitable disaster.

The Roots of an Obsession

During his first stint as NASA Administrator, Dr. James Fletcher was instrumental in winning support for the Space Shuttle program. In 1976, Dr. Fletcher testified before the U.S. Senate that the Shuttle "can launch all of the known payloads in the future" and that the Shuttle flight rate "would go up to 60 per year." For a decade and a half, NASA officials managed the Shuttle program with dogmatic adherence to the two goals put forward by Dr. Fletcher: to establish the Space Shuttle as the nation's only space transportation system and to fly as many Shuttle missions as possible. It took the destruction of the Shuttle Challenger and the loss of seven lives before this mindset could be altered.

The roots of this obsession date back to 1971, when NASA analysts prepared the cost projections that were used to help win approval of the Space Shuttle program. NASA's goal at the time was to demonstrate that the Shuttle could be operated for about ten million dollars per flight, which is roughly $35 million per flight when adjusted for inflation to the late 1980s. By meeting these cost targets, it was argued, the Shuttle would be able to deliver payloads to orbit for about one tenth of the cost of using the expendable rockets that were in operation at the time.

But to achieve these ambitious cost goals, two important economies would have to be realized. First, the substantial fixed cost of maintaining the Shuttle's extensive launch and processing facilities would have to be spread over a large number of missions. Second, the variable costs of operating the Shuttle would need to be reduced by achieving economies of scale in the production of key Shuttle components, such as the external fuel tanks and solid rocket boosters. Neither of these two economies could be achieved, however, unless the Shuttle were flown very frequently. To prove to the Nixon Administration and Congress that these cost goals could be met, NASA was forced to base its Shuttle cost projections on a very high projected flight rate: 67 Shuttle missions per year, or about one flight every five days.

By the mid-1970s, NASA officials were already feeling tremendous pressure to sign up customers and justify the rosy economic forecasts that were used to help sell the Space Shuttle program to the American people. Due to widely publicized difficulties encountered in the development of the Shuttle's main engines and thermal protection tiles, the program was running well behind schedule, and had hence become a vulnerable target for NASA's critics. Funding for the Shuttle program was also sapping away money from NASA's basic space science and applied research activities, raising the question of whether the agency would have any experiments of its own to fly on the Shuttle. To improve chances that the Shuttle would achieve very high flight rates, NASA announced that it would phase out its use of expendable rockets — the Atlas/Centaur, Delta, and Titan — as soon as the Shuttle could meet the nation's demand for launch services. But NASA's hopes of an easy marketing job were dashed by a new and unexpected competitor: a French consortium, *Arianespace*, which pledged to capture a significant share of the communications satellite launch market with its expendable Ariane rocket.

In its efforts to outsell Ariane and fill up the Space Shuttle "manifest" — the official register of scheduled missions — NASA adopted a very aggressive marketing strategy, which included establishment of a *loss leader*, an artificially low Shuttle price for all non-NASA customers. In private industry, such a tactic is often termed as "cutthroat competition," because a company will sometimes undermine its own financial condition to rid the marketplace of unwanted competition. To the U.S. Government, however, this policy merely amounted to the subsidization of another national resource. The effort was also highly successful; by the time the first Shuttle mission was flown in 1981, hundreds of satellites and experiments had been scheduled by the Department of Defense and private companies for launch on some seventy Shuttle missions extending through 1988. The total number of payloads scheduled to be launched on America's expendable launch vehicles numbered fewer than a dozen.

Keeping the Shuttle economically competitive with the Ariane rocket was only one of NASA's many requirements for developing a broad base of Space Transportation System (STS) users. NASA was counting on the allegiance of two groups of "captive" users, customers for whom the Shuttle was the only possible launch system. Fitting into one of these categories were those customers whose payloads were too large to be flown on expendable launch vehicles such as the Ariane. The other group included companies and entrepreneurs planning new and experimental uses of space, such as zero gravity materials processing, which would require manned interaction or other unique attributes of the Space Shuttle. NASA had a particular interest in cultivating this latter segment of the market, because a major selling point for the Space Transportation System was the claim that it would stimulate the development of new space products and applications by the private sector.

To convince companies to use the Shuttle for commercially oriented research, the problem was not competition from Ariane, it was the large up-front costs and long lead-times that investors would have to endure before mature and profitable processes could be developed. To help eliminate these barriers, NASA planners decided that the agency would need to offer space entrepreneurs much more than low Shuttle prices. With help from Congress, NASA established a variety of incentive programs to encourage companies to use the Shuttle for research and development. Through these programs qualified companies would be eligible for free flight time on the Shuttle, access to NASA laboratory facilities, and other types of support.

Accomplishing an early string of successful Space Shuttle missions remained NASA's foremost priority in its strategy to earn the confidence of its space transportation customers. The agency soon realized, however, that selling a product in a competitive environment can be as challenging as making the product work. Prior to the Shuttle program, most of NASA's marketing experience was limited to convincing the White House and U.S. Congress that its projects should be funded. Beyond this, NASA had very little selling to do, since the agency always assumes the role of "the customer" once its programs are approved. Operating the Space Shuttle became the first major endeavor in which NASA did not enjoy the luxury of being the customer. NASA bureaucrats who had avoided the realities of capitalism for their entire careers suddenly found themselves having to convince hard-nosed, profit-seeking businessmen that a ride on the Space Shuttle is worth its cost and risk. As the space agency rapidly discovered, this entails a completely different challenge than dealing with the politics on Capitol Hill.

Customer Services: NASA's Shuttle Salespeople

The NASA people responsible for selling the Space Shuttle operate out of a suite of plain-looking government offices in a federal office building in Washington, D.C. The NASA Headquarters decor is typical for an agency of the U.S. Government, but is bland for a marketing organization whose customers typically spend tens of millions of dollars at a time. After the Space Transportation System became operational, NASA began to seek ways to make its Shuttle marketing effort more glamorous and "customer-friendly." The agency could not do much to improve its office surroundings, but it did change the name of the organization inhabiting these facilities. The Division of Space Transportation Systems Utilization would henceforth be known as *Customer Services,* which NASA felt has a much friendlier ring.

When the Customer Services Division was formed, it was staffed almost entirely by career government employees and was headed by Chester M. Lee, a salty retiree from the U.S. Navy. Captain Lee conducted weekly status meetings in a small conference room whose walls are adorned with dozens of colorful cardboard cut-outs shaped like the Shuttle Orbiter. Each cutout represents a different Shuttle mission, and has in its cut-out cargo bay a miniature cardboard facsimile of what is scheduled to fly on that mission. The only Shuttle missions whose contents are not pictorially displayed are those of the Department of Defense, whose Shuttle payloads are classified "secret" as a matter of policy.

Each cut-out on the manifest, which includes completed Shuttle missions or future flights whose customers have begun making progress payments, is referred to by the Shuttle mission number. The first nine Shuttle missions were designated STS-1 through STS-9, but beginning with the tenth Shuttle flight, NASA adopted a more complex numbering system, perhaps to avoid the connotations of bad luck that might have been associated with a mission designated STS-13. Engineers are not particularly superstitious, but it might be recalled that the Apollo 13 mission was aborted after a near-disastrous on-board explosion occurred en route to the Moon. Under NASA's new Space Shuttle numbering system, the thirteenth mission was instead designated STS 41-G. The "4" corresponds to the last digit of the fiscal year in which the mission was flown, 1984, while the "1" refers to the launch site (1 = the Eastern Test Range, which is better known outside of NASA as the Kennedy Space Center, and 2 = Vandenberg Air Force Base in California). The letter in the code is based on the sequence of the mission in the flights scheduled for that fiscal year; STS 41-G was originally scheduled to be the seventh mission in 1984, following STS 41-A through STS 41-F.

Customer Services is responsible for overseeing virtually every aspect of Shuttle marketing, which ranges from such critical activities as the development of Space Shuttle pricing policies to tasks as mundane as distributing the miniature Shuttle manifest booklets that are periodically published by NASA Headquarters. Customer Services is the first NASA organization that a prospective Shuttle customer calls to obtain information or to reserve a slot on the Shuttle. To get on the official manifest, a customer must complete a NASA "Form 100" and give Customer Services a non-refundable deposit of one hundred thousand dollars, which is eventually credited toward the launch price. This credit can take awhile to be applied, because a Shuttle reservation must typically be made at least four or five years in advance. When NASA published its November 1985 Shuttle manifest, the last schedule to be released prior to the Challenger accident, customer requirements were defined through the end of 1990, and only four payload openings were available on the forty-seven missions firmly scheduled through August 1988.

As NASA's Space Shuttle marketing efforts became more sophisticated, Customer Services expanded its role, gradually taking on the characteristics of a "one stop shop," where a variety of customer needs are met by one organization. Most of Customer Services' efforts are devoted to meeting the requirements of the Shuttle's private sector customers, for whom use of the Shuttle is a key element in a high-stakes business. The typical communications satellite that is launched on the Shuttle carries a price tag of $50 to $100 million and takes three years to build, and will generate as much a one million dollars per *week* in revenue once delivered to orbit. Such a satellite is literally worth more than its weight in gold after it has been successfully launched, but is worthless as long as it sits on the ground. One of Customer Services' most challenging tasks has been to meet its schedule commitments to get such payloads into orbit as quickly as possible. The numerous delays encountered during the first few years of Shuttle operations have greatly complicated the task of maintaining the STS manifest to the satisfaction of NASA's Shuttle users.

To further add to Customer Services' grief, even the normally routine matter of customer billing can become a numerical nightmare. Unlike other transportation systems, where riders can simply pay their fare and get on board, Shuttle customers must make payments to Customer Services at six-month intervals over a three year period before their payloads even get near the launch pad. To further complicate matters, prices are calculated by NASA in *1982 dollars*. The current Shuttle price was established in 1982, and every time a customer makes a payment, that installment is escalated from the 1982 price according to a government inflation index published monthly by the U.S. Bureau of Labor Statistics. When NASA's initial three-year price subsidy ended in 1985, the price for a "dedicated" (full) Shuttle launch was raised from its initial $38 million (in 1982 dollars) to the price established in 1982: approximately $71 million. But a customer paying a "$71 million" bill in escalated 1985, 1986, and 1987 dollars would actually end up paying Customer Services a total of over $80 million per full Shuttle launch.

Since individual commercial payloads generally consume only a small fraction of the space available in the Shuttle's sixty-foot long cargo bay, an additional complication arises: NASA must determine what share of the full Shuttle price each commercial customer must pay. To come up with a figure, Customer

Services uses an arithmetic formula that takes into account the length and the weight of the customer's payload, determining a *load factor* that represents the percentage of the Orbiter's lift capability that the payload requires. To ensure that a full fee is collected even if there is some unused space in the Orbiter cargo bay, NASA tacks an additional 33 1/3% onto the load factor to obtain the *charge factor,* which determines the percentage of the dedicated Shuttle price that a customer must pay. Using this formula, Customer Services collects about fifteen to twenty million dollars for launching a relatively small communications satellite and its upper stage booster rocket on the Shuttle.

To get a satellite or other payload into orbit, a Space Shuttle customer faces a number of expenses in addition to NASA's basic STS use charge. One such cost is incurred for *payload processing and integration*— the preparation of a payload for launch and its installation into the Orbiter cargo bay. Most Shuttle payloads are so complex and delicate that preparation for launch requires several weeks of painstaking, step-by-step mating with the Space Transportation System. Integration of communications satellites is particularly troublesome because they must be physically and electronically linked with upper stage boosters prior to loading into the cargo bay. Many payload processing operations must be carried out in specially equipped "clean rooms" that are extremely expensive to maintain. For a large communications satellite, the total cost of payload processing and integration can be as much as ten million dollars. Add to this the purchase price of an expendable upper stage, which can range from six million dollars for the McDonnell Douglas *Payload Assist Module* to fifty million dollars for Boeing's *Inertial Upper Stage,* and NASA's Space Shuttle price can appear almost small in comparison.

To protect this investment, most Shuttle customers buy optional flight insurance, although rising premiums may soon force many to self-insure. The Space Shuttle was once considered by insurance underwriters to be the most reliable space transportation system in the world, enabling a Shuttle customer to insure a satellite launch for five or six percent of the insured value. But several problems with Shuttle satellite launches in 1984 and 1985 resulted in large increases in premiums, even though the Shuttle itself was not to blame in any of those mishaps. It remains to be seen what long term impact the loss of the Orbiter Challenger will have on the insurance industry, although Shuttle insurance premiums are unlikely to drop to their original low levels for a very long time, if ever. Shuttle customers are not obligated to purchase insurance to protect the Orbiter or its crew, but their payloads must adhere to volumes upon volumes of safety requirements that NASA has established for all items that fly on the Space Transportation System.

NASA's Customer Services Division provides information and assistance to help prospective users of the Space Shuttle obtain all of these services and meet the various requirements for flight on the STS, although NASA's status as a government agency presents some limitations in its dealings with private companies. One constraint upon Customer Services is that NASA cannot offer deferred payment options to Shuttle customers, even though Arianespace offers attractive payment terms to users of the Ariane launch vehicle. Nor can NASA offer to sell launch insurance, another option offered by Arianespace. Customer Services is also prohibited from making creative business arrangements with STS users, such as accepting a percentage of future satellite revenue in lieu of up-front payments for launch services. Despite these constraints, NASA has sustained its efforts to achieve the highest Space Shuttle flight rates possible, aggressively marketing the STS and trying to help Shuttle customers overcome the many technical and economic barriers with which they must contend.

NASA's Shuttle Price War

Since the Shuttle began operating in 1981, the most aggressive element of NASA's STS marketing program has been its Shuttle pricing policy. Even though it costs millions of dollars to fly a typical Space Shuttle payload, NASA gives its commercial and DOD customers a tremendous bargain. Through fiscal 1985, NASA spent approximately six billion dollars on Space Shuttle operations and completed a total of twenty-two missions, for an average cost per flight of over $250 million. This does not include the $18 billion cost of developing the Shuttle and building the first four Orbiters, which, if spread over the entire expected life of the Shuttle system, would probably add another $50 million to the cost of each flight. While this amounts to a total effective cost of $300 million per flight, NASA charged its customers at an average rate of less than $50 million per flight over this period.

NASA was successful in its efforts to establish a low early Shuttle price largely because of expectations that the Shuttle would be inexpensive to operate. One of NASA's key selling points in getting the Shuttle program approved was the agency's argument that the STS would be largely self-supporting; it was in fact hoped that NASA could turn a small profit on Shuttle operations, helping to offset the taxpayer investment in Shuttle development and Orbiter production. To get the Nixon Administration to approve the Space Shuttle program in 1971, NASA made its commitment to the Office of Management and Budget that the operations cost for each Shuttle mission would be no more than $10.5 million, which is about $35 million in inflation-adjusted 1987 dollars. These optimistic projections were then propogated by NASA and its industry Shuttle team for several years. In fact, Shuttle prime contractor Rockwell International subsequently revised its

Shuttle operations cost estimates *downward,* predicting in an October 1973 report that the cost per flight would be $8.5 million, or approximately $28 million in 1987 dollars.

As the Shuttle development program progressed, these operations cost estimates were gradually increased. By 1976, when the Orbiter test vehicle *Enterprise* was unveiled and the first Shuttle prices were established, NASA had revised its Shuttle cost projections upward by nearly fifty percent, to about $18 million per flight in 1975 dollars. This cost projection was used as the basis for establishing the price NASA would charge its Shuttle customers for the first three years of Shuttle operations. Commercial users would pay a rate of $18 million per full Shuttle launch; in 1987 dollars, this is roughly $45 million. The price for the Department of Defense, which would be effective for the first six years of Shuttle operations, was set at $12 million ($30 million in 1987 dollars), a figure one third lower than the commercial price. The official reason given for the lower DOD price was was that NASA expected to obtain offsetting benefits from the Defense Department for Shuttle missions launched from Vandenberg Air Force Base in California. The real reason for the additional price break: the Defense Department had enough political clout to hold NASA to its original low cost estimates.

In view of the fact that the actual cost per Shuttle mission has been at least $250 million per flight, these prices seem absurdly low. The $45 million commercial price and the $30 million DOD charge represent only a small fraction of what it has really cost to fly a typical Shuttle flight over its first three years of operation. The remainder of the cost of each mission has been borne by NASA, meaning America's taxpayers. How did the U.S. Government end up subsidizing each Space Shuttle mission to the tune of over two hundred million dollars per flight, when the system was originally intended to be self-supporting? To understand the initial Space Shuttle prices, it is necessary to consider NASA's objectives and expectations when these rates were established.

NASA's Rationale for Subsidizing the Shuttle

In 1976 NASA estimated that 572 Shuttle missions would be flown over the first twelve years of the system's operation, at a total cost equivalent to $25.4 billion in 1987 dollars, or an average of $45 million per flight. The space agency therefore expected that, by establishing a $45 million price, it would be charging commercial customers the *long run average cost* of flying the Space Shuttle. It was recognized by space agency officials that early operating costs would be much higher than the long run average cost, but it was hoped that maintaining the lower price would stimulate use of the Shuttle, enhancing its long term cost-effectiveness. Moreover, NASA claimed that the U.S. Government would eventually break even or earn a profit, because at the conclusion of three years the Shuttle price could be increased to a figure higher than the long run average cost. NASA thus had a twofold rationale for subsidizing users of the Space Shuttle: it would encourage the use of a national resource, and it would not result in a significant taxpayer burden, based on expectations that Shuttle operating expenses would be low enough to enable the government to recover its investment within a reasonable period of time.

NASA's failure to accurately predict the cost of operating the Shuttle continued throughout the entire Shuttle development period and into the early operations phase. As recently as the summer of 1981, NASA projected average Shuttle operations costs of approximately $90 million per flight, and still harbored hopes of recovering all Shuttle operating costs within a twelve year period. Hence, even after the first Space Shuttle mission had been completed, NASA's estimates of Shuttle operations costs were a factor of three lower than actual operating costs. These 1981 cost estimates were developed to support preparation of NASA's "Phase II" Shuttle pricing policy, which would cover the years 1986 through 1988. Based on its 1981 cost estimates, NASA considered revised commercial prices as low as $75 million per flight.

Before the Phase II price could be established, however, NASA was offered its first realistic glimpse of Shuttle operating costs, which nearly traumatized the agency. After the completion of the second Shuttle test flight in November 1981, NASA Headquarters performed an in-depth study of Shuttle operations costs, and came up with conclusions that nobody inside or outside the agency seemed to expect. The estimate of the long run average cost, over the first twelve years of operations, was revised to over $140 million per flight, a 55% increase over the projection made less than six months earlier. To make matters worse, the costs during the first few years of operations were projected to be even higher. It was estimated, for example, that the cost of each of the five missions scheduled for 1983 would be over *$380 million*. In subsequent years, according to the analysis, costs would decline sharply as a result of higher flight rates and improvements in procedures, but would never fall below $100 million per flight.

These findings raised considerable concern within the space agency, which at the time was negotiating with White House officials and the U.S. Congress for the lowest Phase II price to which the agency could get the rest of the government to agree. To justify a Shuttle operations budget large enough to continue offering large discounts to non-NASA customers, NASA officials maintained their argument that higher Shuttle prices would discourage STS use and reduce the flight rate. This, according to NASA, would result in still higher operating costs per flight, further reducing the Shuttle's attractiveness and its flight rate. In short, high STS prices would disable the STS, imperiling America's investment in a valuable resource and the nation's preeminence in space technology. To support their argument for lower Shuttle prices, NASA officials developed statistics that, in their view, provided a more reasonable basis for calculating the percentage of total STS operating expenses that should be reimbursed to the U.S. Government by Shuttle customers. Shuttle pricing specialists presented analyses that referred to "aggregate cost," "total operations cost," "marginal cost," "additive cost," and "out of pocket cost," precipitating a battle of terminology that even seasoned economists found confusing. To many observers, the equations linking all these terms seemed as complex as those that were used to design the Shuttle Orbiter.

Despite NASA's obvious attempts to use any means possible to justify large subsidies, the agency was apparently successful in convincing the bureaucracy that a low Shuttle price was essential to the system's success. In June 1982 the Phase II price was officially established at $38 million in 1975 dollars, or just over $90 million in 1987 currency. Based on NASA's own cost projections, this represented a continued government subsidy to commercial Shuttle customers of at least $30 million to $60 million per flight; more pessimistic cost estimates suggested that the effective subsidy would be more like $100 million to $150 million per STS mission.

This low price was established despite the active lobbying of France's Arianespace, which argued that a continued STS price subsidy would create unfair competition in the launch vehicle industry. The U.S. Government was not responsive to Arianespace, largely because French and European governments were busy subsidizing Ariane's price as well. In fact, European users of the Ariane rocket were being charged higher prices to use their own system, so discounts could be offered to the American companies that Arianespace wanted to lure away from the Space Shuttle. While Arainespace claimed that these practices were necessitated by the American subsidy of its Space Transportation System, they would not guarantee that Arianespace would end its subsidies if the United States ended its financial aid to Shuttle users. Thus began an unusual competition to see which government could out-subsidize the other, with the goal of winning over customers that would pay only a fraction of the cost of the services they were receiving.

Shortly after the Phase II pricing policy was established, NASA began developing its Phase III policy, which would cover launches beginning in fiscal year 1989. This policy would have to be established by late 1985, when customers for the first Phase III Shuttle launches would begin making progress payments. In presenting its arguments for a second continuation of the Shuttle price subsidy, NASA found more opposition than it had in the previous round, because an emerging domestic launch vehicle industry began to put pressure on the U.S. Government to raise the Shuttle price. During the 1982-1984 time period, several American companies announced their intent to market expendable launch vehicles (ELV's) privately. General Dynamics Corporation, which had no desire to close down its Atlas/Centaur production line just because NASA wanted to fill up its Shuttle manifest, officially began its attempts to market the Atlas/Centaur commercially in early 1983. At about the same time, a start-up firm, Transpace Carriers, began marketing McDonnell Douglas' Delta ELV. Another fledgling firm, Starstruck, Inc., announced it would begin developing a new commercial rocket, the Dolphin, which would utilize an unusual technology for combining liquid and solid fuels.

These companies found it very difficult to sell launch vehicles with their chief competitor — NASA — giving its customers the equivalent of a seventy percent discount. With the support of space commerce advocates interested in seeing the development of a commercial launch industry, these firms stepped up pressure on the U.S. Government to force NASA to base its Shuttle price on *total* STS operating costs. Arguments against continued Shuttle price subsidies appeared to be given a boost by President Reagan's 1983 and 1984 statements in favor of commercial space transportation, and were presented at Congressional hearings in February 1984. The newly formed *Commercial Space Transportation* office within the U.S. Department of Transportation, which was given responsibility for eliminating barriers to commercialization of launch vehicles, became the first U.S. Government agency to officially support a revision in NASA's Shuttle pricing policies.

To counter NASA's arguments that a low Shuttle price would be necessary to keep the Space Transportation System efficient, these groups contended that the Space Shuttle should be used only for those missions that absolutely require the unique capabilities of the Orbiter. One reason the Shuttle's costs are so much higher than anticipated is that the system must do a lot more than simply deliver unmanned cargo into orbit. On every mission the STS must accommodate dozens of experiments developed by scientists all over the world, in addition to the more routine business of launching satellites. Moreover, the Shuttle has the stringent safety and reliability requirements of a manned spacecraft, so construction and operation of Shuttle elements require painstaking and expensive quality control provisions. Had the Shuttle been designed solely to deliver unmanned payloads such as communications satellites to low Earth orbit, its design could have been much simpler and the system would undoubtedly be much less costly to operate than the current configuration.

Some supporters of a competitive launch vehicle industry suggested that NASA adopt a multi-tiered Shuttle pricing formula, extending price subsidies only to organizations using the STS for experimental projects that require manned supervision or other unique Shuttle attributes. Under such a pricing scheme, profit-making firms whose payloads could just as easily be flown on commercial launch vehicles would be charged the full cost of their STS utilization. The additional revenue gained from customers with mature money-making space enterprises would then be used to reduce taxpayer costs, or to help support Shuttle use for scientific purposes or applications aimed at developing new commercial products or processes.

With a number of new groups favoring a significant increase in Shuttle prices, many industry experts predicted that the Phase III Shuttle price would be increased at least fifty percent from the $90 million Phase II price. But despite the arguments in favor of such an increase, the Reagan Administration in 1985 approved a NASA plan for auctioning commercial Shuttle launches to the highest bidder, with a minimum acceptable bid price of about $80 million per dedicated flight. The new procedure, which is effective for Shuttle launches beginning in fiscal 1989, could actually result in Shuttle prices *lower* than the Phase II rates. While an increase over Phase II prices is also possible, the new policy will almost certainly keep the Shuttle price well below actual operating costs. NASA's success in convincing the U.S. Government to limit Shuttle prices, at least through the early 1990s, is probably due in large part to the favorable impression the Space Shuttle has made on the American people. Despite the fact that its economic potential, so great a factor in its selection, has been far from realized, the Space Shuttle remains, by popular consensus, *the* nation's Space Transportation System.

Did NASA's Shuttle Salesmanship Lead to Disaster?

NASA's aggressiveness in trying to attract Shuttle customers raises the question of whether it is appropriate for a government agency, particularly one that is oriented toward research and development, to engage in free market competition. By delegating responsibility for operation of communications satellites to the private sector in the early 1960s, NASA helped foster the development of a large and prosperous industry. In contrast, NASA's Shuttle price subsidies of the 1980s made it nearly impossible for a commercial space transportation industry to emerge in the United States. One must wonder if the space transportation market would be developing in a healthier manner if the benefits of free enterprise were being relied upon, rather than centralized bureaucratic control.

An even more troubling question is whether NASA's eagerness to maintain a high Shuttle flight rate might have contributed to the January 1986 accident that resulted in the loss of Challenger and its crew. NASA was clearly under pressure from 1983 through 1985 to reduce the number of Shuttle launch delays, which had reduced the number of missions completed through fiscal 1985 to twenty-two, out of a planned total of thirty-two. The ill-fated Challenger launch had already been delayed twice, and any further delays would have made it nearly impossible for NASA to achieve its 1986 goal of launching fifteen Shuttle missions, nearly twice the maximum number of missions flown in any previous year.

To outside observers, NASA's plan for fifteen STS launches in 1986 may have seemed very ambitious, but in the context of the space agency's original goals, this was a bare minimum flight rate for maintaining the continuity and economical performance of the system. When the Shuttle program was approved in the early 1970s, NASA's stated objective was to fly sixty-seven STS missions per year by the mid-1980s. As recently as 1981, the official flight rate goal was fifty-five missions per year, also to be achieved by the middle of the decade. NASA's official "STS Traffic Model" was modified in 1982, but even this schedule called for a "high" flight rate of forty missions per year and a "low" rate of twenty-four Shuttle flights per year. It was only after actual Shuttle operating experience was gained in the 1983-84 time period that most NASA officials began to accept the fact that even the 24 flight-per-year goal would be a challenge to accomplish. The 1982 Traffic Model calling for this flight rate included seventeen 1986 missions and twenty-one 1987 flights, leading up to the 24 per year rate from 1988 onward.

In view of these expectations, NASA had already scaled down its hopes considerably by accepting fifteen Shuttle flights as a goal for 1986. To curtail the schedule further would have resulted in additional adverse publicity and a potential loss of Shuttle customers. In the view of many NASA officials, any further reductions in the Shuttle flight rate would have resulted in unacceptably high operating costs per flight, threatening the competitiveness and success of the entire system. To achieve its busy 1986 launch schedule and to have any hope of meeting its goals for the Space Transportation System, NASA officials considered it vitally important to get Challenger off the launch pad as quickly as possible.

Under these pressures and the agency's desire to live up to its STS promises, NASA officials were probably very reluctant to accept an extensive delay in the launch of Challenger. But if pressures to increase the Shuttle flight rate had not been present, would NASA have been more receptive to the engineers who had recommended that Challenger's launch be delayed? One can only speculate as to the answer to this question, but even if it these concerns did not contribute to NASA's decision to send Challenger on its last journey, the prospect of NASA stretching its resources to fly twenty-four or more Shuttle missions per year, every year, with an aging fleet of Orbiters, seems somewhat disconcerting.

Improvements in Shuttle design and launch decision making will hopefully prevent a recurrence of the Challenger tragedy, but NASA's approach to Shuttle marketing must be called into question if the nation is to have safe and effective space transportation. Six weeks after the loss of Challenger, NASA caved in to public pressure and finally endorsed the development of a private U.S. launch vehicle industry. At the same time, however, the agency continued to project Shuttle flight rates of at least 24 missions per year, and failed to offer any indication that it intended to modify its Shuttle pricing policies or offer any substantial support to private launch vehicle companies. It may therefore remain impossible for private firms to offer launch services at a price low enough to compete with the Shuttle.

The short history of the Space Shuttle program suggests that a broader national debate on the future of the Space Transportation System may be in order. The successes of the program have been significant, and for these the employees of NASA and its contractors should be commended. But a number of critical decisions made by NASA in defiance of outside criticism have resulted in an unprecedented crippling of America's space transportation capabilities. NASA should aspire to Shuttle flight rates even more modest than current plans, focusing its marketing efforts on those customers that need to use the Shuttle's unique capabilities. This would probably enhance Shuttle safety and reliability, reduce taxpayer subsidies for Shuttle users, and result in the development of a vigorous commercial space transportation industry. To ease the adjustment NASA would have to make to accept such a change in its approach to marketing the Shuttle, the nation should make it clear that the success of the Space Program will not be judged solely by the number of Space Shuttle missions we can fly. A simple way of transmitting this message to NASA would be to increase funding for some of its long term activities, such as the Space Station, and to invest in developing even better transportation systems than the Shuttle. Our approach to utilizing the Space Shuttle should not be based on fifteen year-old expectations, but on the realities of today's market and a well formulated plan for creating and taking advantage of tomorrow's opportunities.

3

Next on the Agenda: The Space Station

NASA's success in moving the Space Shuttle into its operational phase cleared the way for the space agency to concentrate on its next long range objective: to develop a manned Space Station. Even before the Shuttle had completed its four test flights in 1981 and 1982, NASA Administrator James Beggs publicly pronounced that the Space Station would be "the next logical step" for America's Space Program. This was a clear signal from NASA's top management to the agency's advanced planners and industry supporters: all systems were "Go" for a unified aerospace industry effort to develop a Space Station concept that NASA could sell to the U.S. Government and the American people. But five years after NASA began its all-out campaign to get the program approved, the Space Station still existed only on paper. Its projected cost increased from about $8 billion to over $13 billion, leading NASA officials to talk about scaling back its effort and accepting greater risks to stay within budget. Is history repeating itself, or has America learned enough from its Shuttle woes to successfully deploy the second half of our Gateway to Space?

The Rise and Fall of Skylab

NASA's dream of establishing a permanently manned Space Station in Earth orbit is as old as the agency itself. If not for the political forces that have set the direction of the U.S. Space Program, such a facility might have been built years ago. When NASA was established in 1958, many of the agency's founding fathers believed that the nation's first great space project should have been to develop a manned Space Station. It is ironic that the Space Station was sidetracked by the Moon race of the 1960's, because a station would have been an excellent first step toward visiting the Moon and other planets. The most sensible and pragmatic approach to going to the Moon, as once argued by Wernher von Braun and other prominent experts, was to stage lunar excursions from a Space Station in low Earth orbit. Voyages to the Moon would be carried out by reusable spacecraft called orbital transfer vehicles, which would ferry personnel and supplies between the Space Station and the Moon.

This "Earth-Orbital Rendezvous" scenario was originally proposed several years before the first manned flights to the Moon in 1968 and 1969, but was not adopted. Instead of developing a permanent station and a permanent transportation system capable of reaching the Moon, NASA elected to develop enormous expendable rockets that could reach the Moon in a single shot. NASA selected this "Lunar-Orbital Rendezvous" technique so President Kennedy's goal of landing a man on the Moon by the end of the 1960s could be accomplished. This objective was achieved, but as a result the nation missed an opportunity to establish the permanent manned presence in space that remains so elusive in the late 1980s. Had the alternative Earth-Orbital

Rendezvous method been selected, a large manned Space Station could have been in operation since the early 1970s.

While the goal of establishing a permanent Space Station was rejected, Project Apollo advanced the state of space technology to such a point that the development of such a facility was no longer a great technical challenge. Some of the basic technologies needed to build a Space Station were even available when NASA conducted its first formal Space Station design study in 1961, and in many respects a Space Station might have been an easier goal to accomplish during the 1960s than sending a man to the Moon. The achievability of a manned space base was demonstrated by *Skylab,* the world's first space station. While the glamorous lunar landings were taking place, NASA quickly and quietly designed and built this station out of modified Apollo hardware. Skylab was designed, built, and launched within a period of about five years and at a cost of roughly four billion dollars, less than a tenth of what the Apollo Program had cost.

The Skylab facility was boosted into a 270-mile Earth orbit on May 14, 1973 by an unmanned Saturn V launch vehicle, the huge moon rocket of the Apollo Program. Sixty-three seconds after liftoff, a shield designed to protect Skylab from meteoroids and the heat of the sun deployed inadvertently and was ripped loose. The loss of Skylab's sun shield, which caused additional damage to the station's solar power arrays, threatened to end the Skylab program before any of its objectives could be achieved. Eleven days later, the three-man crew of Skylab I rode a Saturn IB rocket into orbit with the hopes of saving the crippled Skylab facility. By deploying a makeshift sun shield through the Skylab's scientific airlock and by manually adjusting the lab's remaining solar array, the Skylab I crew made it possible for NASA to carry out its three Skylab missions as planned. The Skylab I astronauts remained in orbit for twenty-eight days, performing nearly four hundred hours of scientific activities and doubling the previous record for human stay-time in space. Five weeks after the Skylab I crew returned to Earth, the three astronauts of Skylab II were launched into orbit, beginning a fifty-nine day mission during which over one thousand hours of astronomy, Earth observations, and zero gravity experimentation were performed.

The record-setting achievements of the Skylab II astronauts were in turn surpassed by the crew of Skylab III, who initiated their stay in orbit in November of 1973. The nine astronauts of Skylab I, II, and III dramatically increased our understanding of Earth as well as improving our familiarity with the space environment, and demonstrated the ability of humans to live and work in space for prolonged periods. Many of the experiments conducted by the Skylab astronauts, such as investigations into the effects of weightlessness on materials and processes, set the stage for work performed more than a decade later on the Space Shuttle. The Skylab project was in fact so successful that the on-orbit stay of the third and final Skylab crew was extended for nearly one month, from fifty-eight to eighty-four days. Due to a costly lack of foresight and commitment, however, this crew was destined to be America's last group of Space Station inhabitants for at least twenty years.

NASA had originally planned to use the Space Shuttle to visit and reactivate Skylab as soon as the Space Transportation System became operational. But Skylab's orbit decayed faster than expected and the station plummeted to Earth on July 11, 1979, disintegrating over the southeastern Indian Ocean and a sparsely populated section of western Australia. The unfortunate loss of Skylab less than two years before the first Shuttle mission was due in part to natural occurrences beyond NASA's control. An unexpectedly high level of solar sunspot activity in 1978 and 1979 expanded Earth's atmosphere, accelerating the degradation of Skylab's orbit. But lack of vision by the space agency and lack of support by the U.S. Government also contributed to Skylab's untimely demise. Had a relatively small additional investment been made to increase Skylab's store of station-keeping propellants, the facility could have been raised to an altitude high enough to allow the facility to remain in space until the Shuttle could be used for a rescue mission. Another possible opportunity to save Skylab was lost when the U.S. Government failed to provide NASA with sufficient resources to keep the Shuttle development program on schedule. If Columbia's maiden voyage had been performed in March 1979 as originally planned, prospects for salvaging Skylab would have been far more promising.

Were it not for Skylab's unfortunate fate, NASA could now have a space station even larger than the one the space agency plans to establish in the 1990s. Skylab's main orbital workshop was derived from one of the huge liquid hydrogen tanks of the Saturn V rocket, and provided nearly ten thousand cubic feet of habitable volume. This is more than twice the volume contained by the basic module NASA is designing for its 1990s Space Station. Skylab's main workshop was so large, in fact, that it had a running track along its internal circumference, enabling Skylab crews to enjoy zero-gravity jogging sessions. The habitat and laboratory modules being developed for NASA's new Space Station will never provide such amenities, because they are limited in size to fit into the Shuttle Orbiter cargo bay. It will probably be well into the twenty-first century before any American Space Station inhabitants can experience the roominess enjoyed by the Skylab astronauts in the early 1970s.

The Rebirth of the Space Station

NASA elected not to use the loss of Skylab as an excuse to lobby for a new Space Station program, deciding instead to downplay its Space Station activities until it could regain the public's confidence by successfully completing the development of the Space Shuttle. But as soon as NASA felt this goal had been achieved, it rapidly moved the Space Station to the forefront of the agency's planning. Within a year of Columbia's first test flight, NASA Headquarters established a Space Station Task Force to map out the agency's strategy for getting a Space Station program underway. Seasoned NASA veterans from its field centers across the country were brought to the nation's capital to put together a blueprint for NASA's last great enterprise of the twentieth century.

The first job facing the task force was to assess the political environment in Washington, so NASA could determine how ambitious a program could be sold. A variety of approaches were considered, ranging from offering the president and Congress a two billion dollar "Minimum Space Station," to grand ventures designed to capture the public's imagination. One of the unique features of the formative months of the Space Station program was the emphasis placed on getting private enterprise involved in the project in innovative ways. There were many good reasons to try to get the private sector to lend NASA a financial helping hand, including concern over the federal budget deficit, which was perceived as the greatest threat to Congressional approval of the new program. NASA managers also believed that a commitment to the Space Station from private industry would appeal to President Reagan's belief in the virtues of free enterprise, enhancing prospects for White House approval. At one point in the early planning stages, high level NASA officials considered trying to get a large number of companies to pledge small amounts of money to the new project, as a symbolic show of support. As part of this scheme, which never was implemented, it was even suggested that an ex-astronaut make a series of inspirational speeches to Wall Street financiers to stir their enthusiasm for the Space Station.

These approaches were ultimately ruled out in favor of a more conventional strategy, the awarding of "Phase A" contracts to study the uses and potential benefits of the Space Station. NASA typically awards such study contracts, also known as "concept definition" or "mission analysis" studies, as a first step in getting a major program underway. Studies of this type, which generally last for about one year, are followed by more in-depth "Phase B" definition and design studies and ultimately by "Phase C/D" contracts for final design work and hardware production. In an effort to build a broad base of industry support for the Space Station project, NASA planned to set aside a relatively large pool of money for its initial Phase A studies: six million dollars. NASA also decided to award a larger number of such contracts than usual; at least six different companies would be given a share of this money to help NASA define and sell its new program.

NASA solicited bids for these contracts in June 1982, and eight of the nation's largest aerospace firms responded. Rather than spend the summer trying to select six winners, NASA quickly decided to split its six million dollars eight ways, so all eight companies — Boeing, General Dynamics, Grumman, Lockheed, Martin Marietta, McDonnell Douglas, Rockwell, and TRW — became winners. Starting in September of that year, the eight new Space Station contractors were hired to support NASA as sales agents. In an orientation meeting at NASA Headquarters, representatives from the eight firms were directed to canvass the country in search of other companies, universities, and government agencies that might be convinced to support and eventually use a permanently manned Space Station. All eight contractors hired other companies as subcontractors, some for the specific purpose of identifying potential users of a Space Station. The contractors were cautioned by NASA not to develop detailed designs for the Space Station before its most valuable uses were defined. The space agency wanted only a broad definition of the best Space

Station "architecture" — that facility which would be useful to the greatest number of potential customers.

Through these mission analysis studies, hundreds of potential users of a manned Space Station were contacted, and everything anyone ever thought of doing on a Space Station was documented. When the studies were concluded in April 1983, the contractors presented NASA with over ten thousand pages of results, identifying more than one hundred principal uses for a manned Space Station. Proposed missions for the nation's Space Station were divided into three basic categories, which were defined by NASA at the start of the study: "national security" (military) missions, "commercial" (profit-making) activities, and "science and applications" projects. Based on the data provided by these three groups of potential users, the contractors developed a general Space Station profile for NASA, addressing such issues as the facility's optimal size and location, when it needed to be available, and what types of services it should provide. The contractors also performed preliminary cost estimates, giving NASA an idea of how difficult it would be to sell this new program to the American people.

The mission analysis contractors agreed that the Space Station would provide a number of important scientific benefits. In the station's zero gravity laboratory modules, scientists could conduct experiments too complex or long in duration to be supported by the Space Shuttle. In addition to exploring the effects of prolonged weightlessness on animals and plants, experimenters could develop new materials, such as exotic pharmaceuticals and crystals, that would be impossible or uneconomical to produce under the gravitational conditions of the Earth's surface. The eight contractors also agreed that the Space Station could support space exploration and astronomy by accommodating telescopes and radio antennas, which could be mounted on the core structure of the station by external booms. To support super-sensitive instruments that could be affected by the minute gravitational disturbances created by the activity on board the main Space Station, the contractors recommended that NASA develop separate "free-flying" platforms that would not be manned.

Information provided to the Space Station contractors by the dozens of companies that they contacted indicated that the Space Station application of greatest interest to private industry was commercial materials processing in space. Some of the new products that could be produced in zero gravity, such as high value pharmaceuticals, could be worth hundreds of thousands or even millions of dollars per pound. The Space Station would enable greater quantities of such materials to be manufactured than the Space Shuttle, and at a much lower cost. The mission analysis studies also aroused commercial interest in using the Space Station as a "service station in space," to launch and repair communications satellites. The contractors calculated that reusable orbital transfer vehicles, based at the Space Station, could reduce the cost of delivering communications satellites to geosynchronous orbit by a factor of five. Most of the contractors agreed, however, that the financial barriers to private investment in Space Station utilization would be great, and that the program would have to be well underway before most companies would be willing to risk their capital in ventures requiring use of the Space Station.

The only constituency group targeted by NASA that failed to contribute significantly to the Space Station "wish list" was the Department of Defense, which indicated that a manned Space Station would have relatively few military applications. This did not deter NASA, because the enthusiastic response of private industry, academia, and foreign governments to the Space Station seemed sufficient to justify a "permanent manned presence" in orbit. With the potential benefits of the Space Station identified,

NASA set quickly to work in the spring of 1983 to move into the next phase of its Space Station sales job. The space agency's approach to building support for the Space Station was deliberate but subtle. Within the agency, it was a foregone conclusion that the Space Station would be America's next great space project, but NASA would not present its arguments to the rest of the nation until its plans were very well defined. As recently as June 1983, NASA proclaimed in official Space Station documents that "no approval for a development program is currently being sought."

With the completion of the mission analysis studies, NASA restructured its Space Station Task Force for the job of interpreting the study results and developing an initial Space Station design concept. Despite the objections of many scientists, the members of the task force quickly agreed that the Space Station should be permanently manned from its inception, with an initial crew of at least four to six persons. The mission analysis contractors had been in unanimous agreement on this and most other critical design issues. The station would have to include at least one and preferably two habitat modules containing provisions for eating, sleeping, and recreation. Laboratory modules, similar in size and shape to the habitat modules, would be needed for internal experiments and commercial space processing activities. Space Station supplies and spare parts would be delivered to the station at ninety-day intervals in logistics modules, which would serve as the station's warehouse while on orbit.

NASA's original plans for assembling the Space Station elements resulted in a design known as the *Power Tower*. It consisted of a 400 foot long vertical beam, with four interconnected pressurized modules clustered near the "Earth-pointing" (lower) end of this truss. The station would be powered by solar energy, which would be collected by arrays of solar cells extending perpendicularly from the main truss structure. External instruments and experiments would be mounted on pallets attached to the truss; these and all other attachments would be accessed by mobile robotic arms controlled from within one of the pressurized modules. Berthed at the Space Station would be an unmanned, disk-shaped propulsion stage, the robotically controlled "orbital maneuvering vehicle" (OMV). The OMV would be used to retrieve and repair orbiting satellites and platforms operating in the vicinity of the Space Station. At least one co-orbiting platform would be developed as part of the initial Space Station system, enabling scientists and entrepreneurs to conduct continuous astronomical observations and space processing activities that would be disrupted if conducted on the manned facility.

According to NASA's original plans for the Power Tower, each section of the facility would be delivered to a 270 nautical mile low Earth orbit during a series of seven Space Shuttle missions in 1992, and would be assembled in space. NASA selected a cost target of eight billion dollars (in 1984 dollars) to design and manufacture this initial facility, even though many of the agency's own experts estimated that the station would cost over ten billlion dollars. All parties involved agreed that the station would ultimately cost $20 billion or more, because NASA plans called for the Space Station to be expanded continuously, accommodating a crew of twelve or more persons by the end of the century. During the late 1990s, a fleet of two orbital transfer vehicles (OTV's) would be delivered to orbit for permanent basing at the Space Station, opening up the rest of the solar system for manned and unmanned expeditions that would be staged from low Earth orbit. It was envisioned that by the first decade of the twenty-first century the Space Station would serve as a staging base for manned OTV missions to the Moon and Mars, setting the stage for development of additional manned outposts in space.

As the Space Station Task Force developed its Power Tower design concept, NASA initiated high level political discussions aimed at getting the rest of the United States Government to support the new Space Station project. NASA's best hope for attaining a lasting commitment to the Space Station was for President Reagan to be persuaded to establish development of the station as a major national goal. It was the strength of President Kennedy's commitment that had carried Project Apollo to fruition, and President Nixon's blessing had been the key to getting the Space Shuttle program underway. Reagan had given a mild endorsement to NASA's goal of establishing a permanent manned presence in space during his 1983 State of the Union Address, but this was far shy of the bold statement that would be needed to energize a project of such scale.

In August 1983, space advocates on President Reagan's staff organized a luncheon at the White House that was attended by a dozen influential aerospace executives. During this meeting, which helped crystallize the President's growing interest in the commercial development of space, most of the invited businessmen expressed their support for NASA's Space Station activities. A week later, the president ordered the Senior Interagency Group for Space to conduct a Space Station study of its own, focusing on the issue of "how a manned space station would contribute to the maintenance of U.S. space leadership." Over the ensuing months, rumors began to emerge that NASA and the White House had reached an agreement that would facilitate the Presidential commitment NASA sought. One rumor was that the White House would grant its approval of the Space Station if NASA would agree to limit the annual growth of its budget to one percent above the rate of inflation, funding the Space Station within this constraint.

As the President strode before a joint session of Congress to deliver his 1984 State of the Union Address, Space Station advocates across the country listened intently for for a positive signal. They were not disappointed. The President devoted several minutes of his speech to the Space Program, characterizing space development as one of four great future endeavors for the nation. More to the point, Mr. Reagan directed NASA "to develop a permanently manned space station and to do it within a decade." The President offered his vision of the Space Station as an international program that would permit "quantum leaps" in science and "strengthen peace, build prosperity, and expand freedom." This was the bold statement sought by NASA. In the less than three years that had transpired since its first Space Shuttle mission, the space agency had skillfully mapped out a strategy for its next great adventure, and had obtained a promise from the President of the United States that would keep the space agency in business for decades to come. The Space Station was sold, and appeared well on its way to becoming reality within just a few years.

The Space Station's Turbulent Evolution

In obtaining President Reagan's endorsement of the Space Station, NASA had successfully implemented Theodore Roosevelt's strategy of speaking softly while carrying a big stick. In this case, NASA's big stick was the Space Shuttle, whose achievements in the early 1980s enhanced the space agency's credibility and demonstrated the value of a human presence in space. However, the grace with which NASA eased the Space Station onto the government's list of priority programs belied the severe technical and management problems that were brewing within the agency. On the surface, all seemed well with America's Space Program. The Shuttle Orbiters Columbia and Challenger were in operation, and preparations for the maiden voyage of the new Orbiter Discovery were underway. But unknowingly, NASA had entered its last year of innocence.

After President Reagan's pronouncement, NASA quickly dissolved its Space Station Task Force and replaced it with a permanent Space Station office at the agency's Washington, D.C. headquarters. The first challenge facing the headquarters management team was to resolve a political war festering between the NASA field centers in Houston, Texas and Huntsville, Alabama. The Johnson Space Center in Houston and the Marshall Space Flight Center in Huntsville, rivals since the early 1960s, were embroiled in an all-out battle for control over NASA's next major program. NASA Headquarters tried to resolve this dilemma by dividing Space Station development tasks into four separate "work packages," each of which would be managed by a different NASA center. Through this complex arrangement, the Johnson Space Center, Marshall Space Flight Center, Lewis Research Center (in Cleveland, Ohio), and Goddard Space Flight Center (in Greenbelt, Maryland), were each made responsible for managing their own Space Station offices.

The two largest work packages, each representing about one-third of the Space Station development task, were given to the Houston and Huntsville centers. The Johnson Space Center was granted responsibility for developing the Space Station's main structure, airlocks, and habitat provisions. The Marshall Space Flight Center was charged with developing the station's life support system, propulsion systems, and the "common module" that would serve as the building block for all pressurized living and work areas. The Lewis Research Center was assigned the task of developing the Space Station's electrical power system, and the Goddard Space Flight Center was given the charter to provide the unmanned experiment platforms that would co-orbit with the manned station.

Under this scheme, top level Space Station policies were to be established by a relatively small management team at NASA

Headquarters. Most of the power for making day-to-day decisions, however, was granted to the Johnson Space Center, which was given the additional responsibility of integrating the work of the four field centers. While this may have been a politically acceptable way of designating Johnson Space Center as the "lead center" for Space Station activities, the arrangement made life extremely difficult for the aerospace firms that were competing for Space Station business. Each company was now forced to align itself with one of the four NASA centers, and would need to develop the capability to design and build the specific elements within the work package managed by that particular center. The Space Station arrangement also exemplified the space agency's tendency to establish layer upon layer of bureaucracy, a propensity that was soon to lead to disaster.

NASA formally solicited industry bids for 18-month *Space Station Definition and Preliminary Design* studies in September 1984, eight months after President Reagan's State of the Union commitment. So far, the agency was keeping up with its ambitious schedule, which was aimed at having the Space Station in orbit and operating by 1992. Each of NASA's four Space Station centers was authorized to award two parallel design contracts, which would sustain competition within each work package throughout this phase of the program. In November 1984, ten industry teams, each led by a major aerospace company, submitted proposals to NASA for the eight contracts that would be awarded. In the spring of 1985, the winners were announced. Teams led by Boeing and Martin Marietta won the two contracts to design the station elements within Marshall Space Flight Center's "Work Package 1." Johnson Space Center's "Work Package 2" contracts were awarded to McDonnell Douglas and Rockwell. "Work Package 3" contracts, awarded by the Goddard Space Flight Center, went to teams led by General Electric and RCA. The "Work Package 4" contractors, to be managed by the Lewis Research Center, would be TRW and Rockwell's Rocketdyne division.

The first nine months of these definition and design studies proceeded smoothly. At the end of 1985, with the studies about halfway complete, NASA unveiled a new baseline design, the "Dual Keel" Space Station, which the agency and its contractors deemed preferable to the 1984 Power Tower configuration. But suddenly, 1986 was ushered in with the tragic loss of the Space Shuttle Challenger and its crew. It would turn out to be a year of upheaval for the Space Program, and the questions raised by Challenger would have a severe impact on the Space Station project. Within six months of the Challenger accident, newly reappointed NASA Administrator James Fletcher announced a total reorganization of the agency's Space Station management, designed to consolidate greater control of the program at NASA Headquarters. It was also announced that NASA would seek contractor support for Space Station systems engineering and integration, a role that the management of Johnson Space Center had insisted it was capable of performing without a prime contractor.

As 1986 progressed, the implications of the Challenger accident became more evident, with profound repurcussions on the design and overall philosophy of the Space Station. Suddenly much more sensitive to the issue of safety, members of NASA's astronaut corps began openly wondering how hospitable the Space Station would be to human beings. A major question concerned the wisdom of designing the station to be completely dependent on a Shuttle system whose future was now so uncertain. How long would it take for Shuttle crews to assemble the station if the Shuttle flight rate had to be dramatically reduced? How would the Space Station crew return to Earth if the Shuttle fleet were once again grounded for an extended period?

Questions such as these put NASA in an especially difficult situation. With military and commercial customers abandoning the dormant Shuttle for expendable launch vehicles, Space Station assembly and resupply missions suddenly appeared to offer the strongest remaining rationale for resurrecting the Shuttle fleet. Determined to protect the Shuttle-Space Station combination, NASA made only modest changes in its Dual Keel design for the station. Its components would still be delivered to orbit by the Shuttle, but over a period of a few years rather than several months. A total of seventeen Shuttle flights would be needed to complete the construction, but according to NASA the station could begin operating earlier at a reduced capacity. Still, it appeared almost certain that the 1992 target date to begin station operations would have to slip by at least two or three years.

To ensure that its crew could return to Earth during times of emergency, the Space Station will probably be equipped with a "space lifeboat," perhaps similar to the capsules that returned the Apollo astronauts to Earth after their voyages to the Moon. In another safety-related change, space based orbital transfer vehicles may never be fueled at a manned Space Station. After Challenger, the prospect of storing 200,000 pounds of cryogenic OTV propellants at the Space Station suddenly made NASA planners very nervous. Development of OTV's might therefore require the establishment of separate unmanned platforms for storage and servicing of these vehicles. This would not affect NASA's immediate Space Station plans, however, because delays in the program have already pushed the earliest possible date for initial OTV operations beyond the year 2000.

Despite all of these changes, the realities of the post-Challenger era may eventually result in increases in the size and importance of the manned Space Station. The Challenger accident has reminded us that the launch of any vehicle into space is a risky operation, and in the future we will probably try to minimize the number of manned space launches required to achieve our goals in space. The Space Station will help reduce the need for manned launches from Earth because its crew will be able to perform most of the functions that are presently conducted by humans aboard the Space Shuttle, such as scientific experimentation and the repair of satellites in orbit.

But the Space Station will achieve these goals only if the nation takes full heed of the lessons of Challenger. If NASA is forced to operate on a shoestring budget, the agency will make design compromises that may diminish the safety and performance of the Space Station. As observed by National Space Society president Ben Bova, the United States must provide NASA with sufficient resources to develop the *best* space systems money can buy, and not the *cheapest*. NASA's fiscal year 1987 budget was increased significantly to cover the costs of a replacement Orbiter for Challenger, and this is an encouraging sign. Many U.S. Government officials even seem receptive to the idea of indefinitely sustaining this higher funding level — roughly $10 billion per year — so that NASA is not forced to repeat the mistakes of the Shuttle era as we move into the age of the Space Station. It is up to NASA's leadership to demonstrate that the agency is improving its ways of doing business so that this money can be wisely spent.

II

Paying the Toll

Over the course of the past two decades, the almighty dollar has emerged as the most significant force influencing the U.S. Space Program. In 1961 NASA was given a blank check to put a man on the Moon, but this would turn out to be the first and only time that the space agency could make safe accomplishment of a specific objective its number one priority. Ever since, NASA has been under constant pressure to provide the lowest cost means of getting a job done in space. Our first space station, Skylab, was not built to give mankind a permanent foothold in space — it was an orbiting "lean-to" made out of surplus Apollo hardware. It almost tore itself apart on its way to orbit and it plunged into oblivion long before it could meet its full potential. Only the ingenuity of NASA's engineers and the bravery of the agency's astronauts rescued this mission from complete failure.

The mistakes of Skylab were promptly repeated during the Space Shuttle program. After spending a fortune on six short journeys to the Moon, the U.S. Government decided it would invest only a fraction of this amount in the new transportation system that would give us our only means of access to space for the rest of the century. To acquire even this modest outlay, NASA had to make so many unachievable promises that the Shuttle program was doomed to failure even before it began. Sadly, the same forces appear to be working against NASA's new Space Station program. The space agency has been coerced by budgetary pressure into one design compromise after another, to such a point that NASA's own astronauts have voiced reluctance to live and work in the Space Station facility.

Perhaps the astonishing success of Project Apollo spoiled Americans into believing that a successful Space Program could be sustained without economic sacrifices. Unfortunately, this has not proven to be the case. U.S. citizens seem to take great pride in NASA's accomplishments, but the funding needed to develop the right systems in the right manner simply has not been made available. A return to the free-spending days of Apollo may not be the best solution to our problems, but it certainly appears that NASA is being forced to operate on dangerously slim resources. If we are going to establish a Gateway to Space that provides genuine access to the new frontier, someone is going to have to pay the toll.

To keep this fare as modest as possible, efforts to expand cost-consciousness within NASA and the aerospace industry should continue. We have seen that placing too much emphasis on cost often turns out to be counterproductive and can even be dangerous, but some economies can be achieved without compromising performance and safety. We do not want the cheapest space program money can buy, but on the other hand we should not have to spend more than is necessary to get the job done. NASA has started to experiment with a number of cost-cutting activities that seem to offer opportunities to save large sums of money while still maintaining high technological standards.

Particularly promising are initiatives to make greater use of private enterprise in the development and operation of space systems.

For cost-reduction efforts to be well directed and successful, the nation's space projects may have to be managed differently than they have been in the past. The federal government has maintained nearly total control over the nation's major space programs, even though the economical development of modern frontiers has always required innovative forms of partnership between the public and private sectors. Government organizations are not noted for their efficiency or economy, and the federal budget process makes it difficult for NASA to fund its programs in a cost-effective manner. New procurement methods that feature greater reliance on contractor incentives could help reduce the costs of space projects without forcing us to assume unacceptable risks.

By relinquishing some of its control over space projects and by increasing its reliance on private enterprise, NASA might reduce costs and improve its ability to focus on technology development projects. This is the role NASA was originally intended to play when the space agency was formed in 1958. NASA does not necessarily have to operate every major system it develops. Commercial companies have been operating communications satellites profitably since the early 1960s, and the late 1980s will see the emergence of a new domestic expendable launch vehicle industry. Private companies have shown interest in commercial development and operation of many other types of space systems, and this interest is likely to crystallize when NASA's Space Station is completed in the 1990s. If government and industry can find innovative and effective ways to work together to develop the resources of space, then passage through the Gateway may some day be within the means of anyone who wishes to take the journey.

4

Space Development: Can We Reduce the Costs?

When people become familiar with the aerospace industry, they are often surprised to discover how expensive it is to design, build, and operate space systems. To perform any significant activity in space requires a multimillion dollar investment, so access to space is presently limited to governments and large private institutions. The high cost of space transportation is the most formidable obstacle to the utilization of space resources. The Space Shuttle was supposed to offer the economies necessary to open up the high frontier, but it is now obvious that the STS will fall far short of its financial objectives. As NASA recovers from the Challenger disaster, one of the space agency's greatest imperatives will be to improve the cost-effectiveness of its space systems. A consensus is emerging throughout the aerospace industry that large scale space development will not take place unless the cost of doing business in space can be reduced dramatically.

The Space Shuttle: An Economic Horror Story

The Space Shuttle exemplifies the economic woes that currently plague the nation's Space Program. While NASA's first two dozen Shuttle missions were highly successful from a technical standpoint, they were an economic nightmare, costing the space agency roughly *twenty-five* times what NASA had originally promised Congress the Shuttle would cost to fly. Even after adjustment for inflation, NASA's original Shuttle cost-per-flight projection of $35 million in today's dollars is a far cry from the $250 million or more it actually cost to fly each Shuttle mission over the first five years of STS operations. As a result of these cost overruns, the price NASA collects from its Shuttle customers covers only a fraction of what it really costs to deliver their spacecraft and experiments to orbit. The agency has therefore had to allocate nearly one-third of its annual budget to the routine task of operating the Shuttle, greatly reducing the amount of money available to fund the research and development projects whose pursuit is supposed to be NASA's primary objective.

The fundamental problem with the Space Shuttle is that it cannot be operated at the rates necessary to make the system economical. In arguing for development of a reusable launch vehicle in 1971, NASA predicted that the Shuttle would be flown sixty-seven times per year by the mid-1980s. If the Shuttle were currently being flown at such a rate, the system might be coming close to achieving its financial goals. The one billion dollars in fixed costs NASA incurs each year to maintain its Shuttle launch facilities, if spread over sixty or more missions, would reduce the effective cost per flight by as much as $100 million. In addition, higher flight rates would result in economies of scale in the manufacture of many of the Shuttle's main components. If Shuttle elements such as the external fuel tank and solid rocket boosters could be manufactured in quantities ten times greater than current rates, variable costs could be reduced by as much as $50 million per flight. Such reductions in fixed and variable costs per flight would result in a total reduction of about threefold in the cost of each Space Shuttle mission.

Unfortunately, the prospect of sixty-seven Shuttle flights per year appears nowhere in sight. NASA's efforts to increase the Shuttle flight rate to fifteen missions in 1986 met with disaster one minute into the second flight of the year. The Challenger accident demonstrated that NASA's STS resources are far too limited to support high flight rates, and that the Shuttle itself is too complex to ever be flown as routinely as an airliner. To fly sixty-seven Shuttle missions per year, if this were technologically feasible, would require an STS support system much more extensive than that which NASA presently sustains. A fleet of at least twelve Shuttle Orbiters would be needed, verus the four Orbiter fleet that NASA will have when Challenger's replacement is commissioned. These twelve vehicles would have to be operated from six launch pads and supported by three times the manpower and production capacity currently available.

Even if such a massive operation could be managed, it is questionable whether it would be fully utilized. If every commercial communications satellite in the world were launched on the Shuttle and every space scientist got to fly his or her dream payload on the STS, it is unlikely that more than thirty civilian missions would be flown each year. If military uses of space were to expand substantially, the Department of Defense might add another ten missions per year to the Shuttle manifest. It is very difficult to imagine a space development scenario that would require the sixty-seven Shuttle flights per year predicted by NASA. In their eagerness to justify development of the Shuttle, space agency officials not only underestimated the system's complexity — they also made unrealistically high projections of the demand for Shuttle launch services.

This is not an unusual phenomenon within the aerospace industry. The way space programs are currently managed, the first step in the development of a new system is for NASA and its contractors to convince the public, the President, and Congress that its projects are worthwhile. To accomplish this, NASA and its industry supporters have a tremendous incentive to underestimate costs and to make optimistic projections of demand. Once a program is approved, however, the contractor's incentives to achieve its stated cost goals often disappear. Until very recently, most major contracts awarded by NASA were *cost-plus,* meaning the contractors earned a profit no matter how much their projects eventually ended up costing. Since profits are generally a fixed percentage of sales, cost overruns while developing a product often resulted in *greater* earnings for the companies involved.

Lately, cost-plus contracting has given way in many cases to other forms of procurement that provide contractors with greater incentives to control development and production costs. These include *fixed price incentive* contracts, which require the contractor to pay a percentage of any cost overruns, and *cost plus award fee* contracts, which cover overruns with taxpayer money but provide the contractor with financial bonuses if designated cost targets are met. These alternative forms of contracting make it difficult for contractors to underestimate costs with the idea of billing the government for unexpected overrruns once their programs are underway. However, they do not induce contractors to identify challenging ways to reduce costs during the early design phases of programs, when the economics of a project is typically determined. In fact, by penalizing the contractor for failing to achieve cost targets, these types of contracts may actually encourage contractors to "play it safe" by offering inflated cost projections.

Another major failure of the current procurement practices is that they do not motivate contractors to accurately predict the costs of *operating* the systems they develop. For nearly all space programs, the contractors' responsibilities end when their products are turned over to NASA. Space agency officials grew accustomed to maintaining total control over the operation of launch systems and associated facilities during the Apollo era, and appear intent on maintaining this role throughout the Shuttle and Space Station programs. All of these features of the government procurement process have combined to short-circuit the natural regulatory features of free enterprise, which encourage the providers of goods and services to make realistic market projections and to find innovative ways to offer their customers the lowest possible prices.

Learning From the Shuttle Experience

NASA's inability to meet its Shuttle usage projections or cost goals raises two fundamental questions regarding the role of economics in the future development of space. First, what *realistic* reductions in the costs of space systems can be expected to result from the expansion of our space activities? Second, how can the current way of conducting business in the aerospace industry be modified so that costs can be reduced in areas where dramatic increases in demand are not likely to provide new economies of scale? If NASA and the aerospace industry are to adequately answer these questions, new emphasis will have to be placed on economics as a fundamental factor in the space planning process. Once a strategy for space development that takes into account economic realities is mapped out, new ways of implementing these programs will have to be found.

Achievement of these goals will require fundamental

changes in the culture and attitudes of many people within the Space Program. When President Kennedy committed the United States to landing a man on the Moon by the end of the 1960s, money was the least of NASA's concerns. The space agency had a clear mandate to put a man on the Moon as soon as possible, no matter what the cost. Congressional budget-trimming, now a standard feature of NASA's fiscal process, was unknown to the Space Program until Project Apollo was well underway. By 1965, the seven year-old space agency had grown by a factor of ten, and was being funded at a level equivalent to over $20 billion per year in 1987 dollars. Uncle Sam's generosity to NASA enabled the United States to fulfill President Kennedy's dream, but it also had some unintended long term results. By giving the space agency a "blank check" during its formative years, the Government helped breed a culture that regarded technical achievement as its most important objective and that considered cost to be a relatively insignificant consideration.

But as the war in Vietnam wore on and the costs of America's new social programs escalated, the public's attitudes toward the Space Program changed. The American people realized that the Space Program had provided some exciting moments, but at a cost of tens of billions of dollars. With little apparent material gain from this investment, questions were raised about the value of space exploration. To adapt to the public's shifting attitudes, NASA realized that its programs would need to offer a more tangible return on the tax dollars they consumed. Thus was born the Space Shuttle, which would be developed for a quarter of the cost of Project Apollo and which would epitomize the nation's new emphasis on practical uses for outer space. The Shuttle's actual performance, however, has shown that economic progress can be an elusive goal. Despite the importance attached to economics by NASA's leadership, the Shuttle has failed to meet virtually every one of its economic objectives.

Understanding this failure is key to achieving success in future attempts to reduce the costs of space development. An obvious explanation for the Shuttle's economic woes is the carry-over of practices from the Apollo era. The 1960s way of doing business was ingrained into personnel at all levels of NASA's organization and throughout the aerospace industry. As the Apollo program was phased out, a tremendous downturn in the aerospace industry occurred. Jobs in the industry became scarce, and those people who were able to keep their positions were predominantly those who possessed the most seniority. While these workers were more experienced, their ideas were generally more established than those of the younger personnel who were forced to abandon their space-related jobs. Even today, many of the space project managers of the 1960s are still in charge, running their programs the same way they did twenty years ago. Under these circumstances, a sudden transition from the free-spending Apollo era to an age of economy was virtually impossible to achieve.

In fairness to NASA and its contractors, it should be pointed out that high technology projects are generally expensive. The research and development process requires the employment of large numbers of highly trained experts, and hence incurs heavy labor costs. A single engineering drawing can require as many as five hundred man-hours of design and related analysis, and thousands of such drawings might be needed for the design of one system. Workers in high technology fields must have access to expensive state-of-the-art equipment such as computers and laboratory facilities, so capital costs are also high. Once a system is designed, it frequently will fail to function the first time it is tried, so after tens of thousands of hours of expensive testing it may have to be substantially redesigned. When the final design for a high technology product is finally completed, extensive factory tooling and costly materials may be required for its manufacture.

These requirements illustrate why the process of developing an advanced technology product can cost billions of dollars. Even in commercial industries, design and development of one new product, such as an automobile or a passenger aircraft, can cost a billion dollars or more. But in a commercial business environment, the research and development phase is usually followed by the mass production of hundreds or thousands of articles. Upfront cost can therefore be amortized over a large number of units, reducing the effective cost per article. In the space world, a massive development activity often results in the manufacture of only a handful of products. The STS is an excellent example; after the $10 billion Space Shuttle research and development effort, only four operational Orbiters were produced.

Recognizing that significant costs would be unavoidable, it was logical for NASA and its contractors to generate optimistic projections of Space Shuttle utilization. From the beginning of the Shuttle program, everyone knew it would be an expensive research and development effort, and projections of the fixed costs of operating the Shuttle indicated that a vast support system would have to be maintained. An economic argument for development of the Shuttle could only be sustained under a highly optimistic demand scenario that would spread these costs over a large number of missions. NASA hence convinced itself and the outside world that its new transportation system would be used so frequently that its production and launch facilities would be operated at near-commercial scale and efficiency. Had NASA performed studies during the 1970s on the effects of lower flight rates on STS operating costs, the ultimate economic performance of the Shuttle might have been predicted far in advance.

Despite these lessons, NASA may now be on the verge of repeating the same mistakes that led to unexpectedly high Shuttle costs. In late 1985, NASA and the Department of Defense initiated extensive "Space Transportation Architecture Studies" (STAS), with the goal of identifying futuristic space vehicles that can reduce the costs of space transportation by a factor of five to ten. These cost reductions are supposed to be achieved by developing newer, more efficient technologies and production techniques, but under current budget constraints it would be difficult to accomplish any major advances in technology. Sensitive to the political problems that would be created if they advocated large increases in research and development spending, the STAS study sponsors based much of their hopes for improved operating economy on predictions of extremely large numbers of future missions. At one point during their study, the STAS contractors assumed that fifty or more satellites will be launched into space or serviced in orbit each year by the end of the century. This seems quite optimistic, since the Space Shuttle, the French Ariane rocket, and American expendable launch vehicles have been launching and servicing a total of only ten to twenty spacecraft per year during the 1980s.

If projections of demand for future space systems are not reduced to reasonable levels, NASA may once again find itself struggling to meet unachievable cost objectives. It is possible that the next twenty years will see a dramatic increase in U.S. space activity. If the nation elects to deploy elements of the Strategic Defense Initiative (SDI), hundreds of space launches and a variety of orbiting space facilities may be required to meet SDI objectives. Alternatively, the U.S. could embark upon a very ambitious civilian space effort, such as the one endorsed in 1986 by President Reagan's National Commission on Space. But in the absence of a commitment to either of these scenarios, we cannot expect increases in demand to result in the needed reductions in the costs of space development. Before approving the development of any new system, we must be prepared to accept the economic consequences of low usage rates.

Fortunately, there are ways of reducing space development costs that do not require massive expansion of space activities. One method is to develop new, more efficient technologies. This was the stated objective of the STAS approach to cost reduction, and is the approach most frequently adopted by technical specialists. Options assessed through the STAS studies include development of longer lasting and more fuel-efficient propulsion systems, greater use of automation in factory production, and simplification of procedures for operating space systems. A totally different approach to reducing costs — one that is not as popular within the aerospace industry — is to implement sweeping changes in the way space systems are funded and managed. Costs can be reduced by changing the government contract procurement process, streamlining design, test, and evaluation procedures, and establishing cost as a primary consideration in the trade studies that determine how space systems are designed and developed.

The technological approach to cost reduction has traditionally been favored within the aerospace industry, for obvious reasons. New technologies are exciting, and programs to develop advanced technologies create large numbers of jobs and large profits. Technology development is also non-threatening to conventional ways of doing business, to which aerospace employees have grown accustomed over the past thirty years. But these are the very drawbacks of technological solutions; they require large and sustained up-front investments, they entail risks, and they do not help transform the space industry culture into one that is inherently economically efficient. Moreover, at the present stage of space development, which is still essentially in its infancy, the options for dramatic cost improvements through use of new technologies are relatively few. When we succeed in moving a significant part of our space activities into orbit on a permanent ongoing basis, many new opportunities for utilizing cost-reducing technologies, such as those employed in extraterrestrial resource utilization, will emerge.

Changing the Way We Do Business in Space

For NASA and its contractors to succeed in reducing the costs of space development within the next few years, they must revive the cultural transition that was unsuccessfully attempted after the Apollo program was terminated. NASA is showing some signs of renewing these efforts, although it will be several years before the success of these endeavors can be judged. The Space Station Program offers a prominent example of increased cost consciousness. This project will represent the space agency's first major attempt at "design-to-cost," a technique employed by the Department of Defense in recent years to help control the costs of new weapons systems. In a design-to-cost project, achievement of specific cost targets is the overriding program goal, and is placed in importance above all but a few fundamental technical requirements. NASA established a "not-to-exceed" goal of eight billion dollars (in 1984 dollars) as a guideline for its Space Station Definition and Preliminary Design studies, and has stated its willingness to compromise on a number of items on its "wish list" of Space Station functions if necessary to meet this budgetary target.

To get the most out of its Space Station investment, NASA has declared its intent to make a broad reexamination of the Space Program culture. The agency hopes to save hundreds of millions of dollars by raising awareness of cost-related issues at all levels of the project's organization and by making fundamental changes in design, development, and production procedures. Interestingly, use of new technology is not a major feature of NASA's cost reduction strategy. The agency expects to cut costs by using new technologies in its management of the Space Station program, such as through its recent implementation of a state-of-the-art computerized information system, but the technology of the station itself is not being counted on to offer many opportunities for economizing. In fact, NASA and its contractors hope to control program costs by using *existing* technology and hardware whenever possible. In an effort to contain up-front costs, advanced technologies are being considered for the Space Station only in cases where there is reason to believe that they will repay their investment by resulting in significant reductions in operations costs.

NASA's approach to design-to-cost emphasizes the need for fundamental changes in attitude among participants in the Space Station program, so that cost is recognized as the second most important design consideration, after safety. All Space Station contractors have been required to justify their major design decisions with supporting cost estimates, to show that unnecessarily expensive design options are not selected. But many industry experts remain skeptical about NASA's prospects for achieving its Space Station economic objectives. By the time NASA's definition and preliminary design studies were completed in early 1987, the agency realized that its initial Space Station concept would probably cost at least fifty percent more than its $8 billion target, even in a design-to-cost environment. Consequently, NASA was forced to scale down the size and scope of the planned station, beginning its familiar process of "cutting corners" to reduce costs even before the start of Space Station full scale development.

Many experts in aerospace economics believe that substantial cost reductions will be achieved in future programs only if fundamental changes are made in the way government and industry work together on space projects. This represents an area of potential cost cutting that NASA has not shown an interest in exploring as part of its Space Station activities. Despite its stated commitment to reducing the costs of this program, NASA appears to be planning a Space Station procurement strategy that is essentially the same as that which was used to acquire and manage other major programs, including Project Apollo and the Space Shuttle. An alternative approach would be for NASA to establish a framework for providing government support to companies interested in sponsoring commercial endeavors to develop Space Station elements. Several innovative forms of government-industry partnership have been employed in the past to increase the efficiency or control the costs of space projects, and such arrangements could also be applied to the Space Station Program.

An example of innovative government-industry partnership is NASA's Joint-Endeavor program, which supports companies involved in microgravity processing. Through Joint-Endeavor Agreements, which NASA has signed with several companies, NASA can offer free Space Shuttle flights and other incentives to reduce the costs and risks incurred by companies investing in materials processing technologies. The first and most well known of these agreements, which NASA signed with McDonnell Douglas in January 1980, established a unique government-industry team with the common goal of manufacturing commercial quantities of ultra-pure pharmaceutical products in space. As its contribution to the partnership, McDonnell Douglas invested tens of millions of dollars of its own capital in materials processing research and development. The firm used part of this investment to develop an experimental production facility that was flown on several Space Shuttle missions. Through the Joint Endeavor Agreement, McDonnell Douglas developed this equipment for a fraction of what it would have cost if it had been created through conventional means of government procurement.

A number of companies have shown an interest in utilizing similar agreements to get involved in the Space Station Program. Such approaches could almost certainly be utilized for development of selected elements of the Space Station, but NASA officials seem to prefer the use of standard aerospace contracts to get the Space Station established. The Joint-Endeavor with McDonnell Douglas was favored by the space agency because it supported a major NASA objective: promoting use of the Space Shuttle. If employed in the development of the Space Station, however, such agreements could reduce NASA's control over the design of the facility and diminish the agency's role in Space Station operations. While NASA officials have stated their interest in Space Station commercialization, the agency personnel who develop commercial policies and administer Joint Endeavor Agreements have had almost no contact with the people managing the Space Station Program.

It is obvious that employing the benefits of private initiative to reduce the cost of the Space Station is a relatively low NASA priority. The space agency has shown a similar disinterest in promoting commercialization in another area that could be threatening to its control over the space business: expendable launch vehicles. If the Space Station and other future programs are to be developed in the manner that is most cost-effective and consistent with the interests of the United States, NASA will have to give up some of its control over the way space systems are designed and operated. This would not necessarily represent a reduction in the space agency's size or status, since NASA would still maintain a significant role in assuring the safety and technical integrity of all space systems. More important, if some of the more routine tasks involved in the development of space could be carried out with greater autonomy by the private sector, NASA could use more of its resources to perform the exciting, far-out activities the agency was created to pursue. In its efforts to reduce costs and enhance America's access to space, NASA should pay greater heed to the economic and social forces that have successfully moved technology forward and pushed frontiers outward since the start of the Industrial Revolution.

5

The Commercial Development of Space

As the harsh economic realities of space development have set in, the need to find alternative ways of financing space projects has gained new urgency. If space remains an arena of activity dominated by government, the future course of space development will be buffeted by unpredictable political forces, and private initiative will play a minimal role in the expansion of the human presence beyond Earth. This is not the way new frontiers have been developed in the past. The North American continent was explored in the fifteenth and sixteenth centuries by expeditions carried out on behalf of European royalty, but large scale settlement of the New World did not occur until a diverse assortment of private groups traversed the Atlantic on their own initiative. Similarly, the U.S. Government used railroad subsidies, the Homestead Act, and other measures to promote the movement of the American population westward, but these activities would have been in vain had people not been willing to take individual risks and invest private resources in populating the frontier states. While government resources are often needed to initiate exploration, permanent human settlement of new territories ultimately requires widespread initiative on the part of individuals or privately-run institutions. Use of government resources to promote private space initiatives may represent our best hope for achieving an economical and effective national space effort.

Space Commercialization in Perspective

Due to the high cost of space exploration and utilization, the public sector will have to dedicate more resources to making space accessible than have been required in the past for the settlement of other frontiers. Once government has established a large space infrastructure, economic factors will probably limit private access to space to large companies for an additional period of several decades. Only after "space industrialization" has been in progress for some time will space become truly accessible to settlement by individuals. Since government cannot lose sight of one ultimate objective — to let the settlement of space by free individuals follow its own logical course — the U.S. Space Program should have as a major goal the establishment of conditions that make private investment in space attractive. *Space commercialization* is therefore becoming regarded as one of the most important forces in the development of the resources of space.

The idea of venturing into space on a commercial basis may seem new, but the National Aeronautics and Space Act that created NASA in 1958 clearly charges the space agency with the task of promoting the commercial development of space. However, this remained a relatively obscure NASA objective for many years. During the first decade of NASA's existence, the glamorous Apollo program and the goal of beating the Soviet Union to the Moon dominated the space agency's budget and the public's imagination. It was not until the early 1970s that NASA and the American people rededicated the Space Program to the goal of promoting commercial enterprises in space. The selection of the Space Shuttle as NASA's first major post-Apollo program reflected the nation's new pragmatic approach to the development of space resources.

After a decade of development and several years of operation, the Space Shuttle has shown mixed success as a vehicle for enhancing industry access to space. Approval of the Shuttle program in 1972 was an important early step in helping to convince private industry that the U.S. Government is committed to the commercial development of space. During its operational phase, the Space Transportation System has become an important technological resource for assisting companies that want to do business in space. On the other hand, the complexity and cost of operating the Shuttle have limited the accessibility of space and consumed NASA funds that might otherwise have been available to provide other means of support to industry. As a result, NASA has had to develop a variety of commercially-oriented programs to encourage entrepreneurs to invest private capital in the development of space.

NASA's various efforts to raise awareness of commercial space opportunities have shown a noticeable impact on private industry. Over the course of the 1980s, it has become increasingly fashionable in financial circles to have a space investment or two in one's portfolio. High-priced conferences and seminars touting the profit potential of space enterprises have become commonplace, and tens of millions of dollars have flowed out of investors' pockets and into new "space companies" like Orbital Sciences Corporation and Starstruck, Incorporated. For many entrepreneurs, the infectious enthusiasm of space business advocates and the glamor of space development have proven difficult to resist. But it remains to be seen whether large scale space commercialization will happen within this century. Private industry will undoubtedly play a leading role in the long term development of space, but there are still formidable technological, economic, and political risks facing would-be space tycoons at this early stage of the space age. Identifying the few good space investments out of the many that are being advertised is a challenging, high stakes game that should be played only by those with a commanding knowledge of this complex industry.

The first rule of space entrepreneurship is: *be prepared to spend lots of time and money before seeing a profit.* The typical commercial space enterprise requires an investment of anywhere from ten million to one hundred million dollars, and will not begin generating revenue until five to ten years after initiation of the project. Investors looking for a quick return would be better advised to keep their money on Earth. A second rule that should be learned is: *become familiar with the aerospace industry and culture*. People in the space world speak a unique language, one that is embellished with technical jargon and littered with confusing acronyms. Learning to communicate is an important key to establishing credibility in the field. A third recommendation: *learn how to do business with the government.* Since the government has traditionally been the largest customer in the space marketplace, its influence is pervasive. While a space venture without government support is doomed to failure, the blessing of the government can be very helpful. Many government resources are available to space entrepreneurs for the asking, and can be used to reduce investment requirements substantially. Fourth and finally: *don't abandon good business sense for the sake of getting in on something exciting*. There is probably no such thing as a low-risk space investment in today's environment. Sound technical and financial expertise are an absolute requirement to separate the few reasonable high-risk investment opportunities from the many that are absurd.

The peculiar requirements for doing business in space have thus far limited the number of successful space entrepreneurs to a select few. Nearly all commercial space ventures that have been attempted or proposed fall into three general categories: satellite operations (including communications and remote sensing of Earth's environment and resources), materials processing in space, and space transportation services. Amid all the hype concerning the tens of billions of dollars that many "experts" project will be earned in these fields, only one of these space applications, satellite operations, represents a mature, profit-making use of the space environment. The profit potential of the other areas of investment, although it may indeed be great, remains to be demonstrated.

Profit by Satellite

It should not be surprising that satellite communications was the first profitable use of space, because it is a concept that is far older than the space age. The idea of utilizing Earth-orbiting satellites for communications in fact predates NASA by nearly one hundred years. In 1869, novelist Edward Everett Hale published *Brick Moon,* a tale about a group of people who were accidentally catapulted into space inside a large brick sphere that was intended to serve as a navigational communications link. Had Mr. Hale known that the cost of launching a single brick into geosynchronous orbit would some day cost approximately one hundred thousand dollars, he might have selected a different construction material for his satellite. The relevance of his basic concept to the space age, however, is fascinating nonetheless.

The first serious proposal for launching communications

satellites was offered by Arthur C. Clarke in the October 1945 issue of *Wireless World,* a scientific journal. Clarke, who gained far greater fame a generation later as the author of *2001: A Space Odyssey,* performed mathematical calculations to show that objects placed in orbit 22,300 miles above the Earth would remain above a fixed location on the planet. Clarke hypothesized that a satellite placed in such a geosynchronous orbit could be used to relay signals almost instantaneously from one point on Earth to another. Satellite communications was turned into reality a decade later by the Soviet launch of Sputnik, mankind's first artificial satellite. Soon after the Soviet feat, satellite communications became one of NASA's earliest areas of emphasis. Today, dozens of satellites routinely beam information from the geosynchronous "Clarke's Belt" and many other altitudes, performing an enormous variety of civilian and military functions.

From the start, NASA viewed space communications as a field whose widespread application would logically be implemented by private industry. NASA officials made statements supporting this proposition as early as 1960. President John F. Kennedy formally expressed the government's advocacy of private ownership and operation of communications satellites in his July 1961 space policy pronouncement. The Kennedy Administration's support resulted in the Communications Satellite Act of 1962, which created the Communications Satellite Corporation, a unique public/private company. According to the provisions of this act, *Comsat* would effectively be granted a monopoly in the intercontinental satellite communications industry, with the goal of establishing a worldwide communications satellite system. To help assure that the activities of the new corporation would be in the nation's best interests, the Comsat Act stipulated that three members of Comsat's board of directors would be appointed by the President and approved by the United States Senate.

In addition to creating the world's first commercial space entity, the Comsat Act itemized a number of specific NASA responsibilities for supporting the new satellite communications industry. NASA was charged with the task of cooperating with Comsat on research and development related to satellite communications, and was made responsible for providing the launch services and associated support required for the emplacement and operation of Comsat's satellites. By permitting NASA to use taxpayer money to support a particular private venture, these provisions of the Comsat Act established a direct precedent for many of the space agency projects that, twenty-five years later, are supporting a variety of commercial space enterprises.

NASA's support for the new satellite communications industry was substantial and effective. In the years 1962 through 1964, the agency's annual budget requests for satellite communications research increased by over fifty percent, to an amount equivalent to roughly $250 million in 1987 dollars. By comparison, NASA's current level of funding for materials processing research is less than one tenth of this amount. Government assistance helped enable Comsat's international spinoff company, *Intelsat,* to launch the world's first commercial communications satellite, Intelsat I, on April 6, 1965. "Early Bird," as the satellite was nicknamed, weighed eighty-four pounds and was capable of transmitting 240 simultaneous telephone conversations or one black-and-white television program. The launch of Early Bird marked the beginning of space commercialization and the first widespread benefit of space to the global population at large.

In the first two decades following the launch of Early Bird, the growth of the satellite telecommunications industry and advancements in communications satellite technology were staggering. By 1985, communications satellites were generating $3 billion a year in revenue from the transmission of television and radio broadcasts, telephone conversations, electronic mail, and business information. In these twenty years, Early Bird was replaced by five successive generations of Intelsat satellites. The most recent of these, Intelsat VI, each weigh over 3,700 pounds and can transmit 33,000 simultaneous telephone conversations and four full-color television channels. According to industry estimates, a total of about $8 billion will have been invested industry-wide in satellites alone (excluding launch vehicles and ground systems) by the year 2000.

The unique and economical services that communications satellites can provide have enabled the industry to grow rapidly despite the significant costs involved in the design, development, and launch of such spacecraft. The cost of building a typical communications satellite can range from $50 million to $100 million, and larger satellites can be even more expensive. Intelsat VI, the most expensive commercial communications satellite, costs about $150 million per copy. Ironically, getting a satellite into orbit is almost equally expensive. The smallest satellites can get a ride on an expendable rocket for about $40 million, while launch of a larger communications satellite can cost over $100 million. But as considerable as these costs are, the investment in a communications satellite can be quickly recovered with the enormous amount of revenue these systems can generate. Annual revenue from one communications satellite is generally at least $15 million and can be as high as $50 million per year, and this level of cash flow can normally be sustained over a ten-year satellite operating life. Nearly all of this money can be applied to investment recovery and profit, because ground system operations for one satellite typically cost only about $3 million per year, or between five and ten percent of operating revenue.

The growth of the satellite communications industry has slowed in recent years, due to the overcrowding of satellites in geosynchronous orbit and saturation of the market for satellite services. Teleconferencing, which a few years ago was expected to become a major new area of satellite communications growth, has not met its initial expectations, primarily because businessmen would rather travel than watch each other on video screens. Another disappointment is the once highly touted market for direct broadcast satellites, which enthusiasts claimed would beam television programs into millions of homes by the late 1980s. With widespread availability of cable television and home receivers that can tune in to conventional satellites, it is now doubtful that a significant direct broadcast satellite industry will ever develop.

Over the next several years, the current communications satellite industry business base will become increasingly vulnerable to competition from the rapidly-advancing technology of fiber optics. Transatlantic optical cables that are currently being developed could offer an economical alternative to the use of satellites for intercontinental communications, eroding demand for the primary commercial application of these satellites. If the satellite communications industry is to sustain its growth, it will have to develop new and unique uses to augment current applications. One application that is a leading candidate to spur future growth is mobile communications. Princeton-based *Geostar, Inc.*, an entrepreneurial outfit founded in 1984, has raised over ten million dollars for development of a completely portable communications system utilizing satellites. The system is intended to enable customers to use relatively inexpensive transceivers to locate their precise geographical positions or transmit digital messages to other customers worldwide. Geostar's venture may ultimately set the stage for the commercial development of very large communications platforms, which could be used for such futuristic applications as "wristwatch telephones."

The communications satellite success story, however, has yet to be repeated for any of the other commercial space industries that are frequently cited as areas of similar investment opportunity. The only other space application to generate significant profits during the first twenty years of the commercial space era has been remote sensing, which employs satellites in low orbits to monitor or photograph the Earth's surface or environmental conditions. While remote sensing satellites have been used since the early years of the Space Program for weather forecasting, the most promising commercial application of this technology is in providing pictures of the Earth's surface to identify valuable resources. The five *Landsat* satellites launched by NASA since the early 1970s have revealed previously unknown lakes and islands, and have been used to map routes for railroads and utility lines. The most extensive commercial uses of Landsat have been in mineral exploration and agriculture, generating tens of millions of dollars in revenue for such services as oil exploration, searches for mineral deposits, and crop forecasting.

Despite these valuable applications, remote sensing seems unlikely to develop into an industry that will rival the satellite communications industry in size or significance. Some users of Landsat data claim that these satellites have already provided more information than could possibly be analyzed and utilized within the next ten years. Owing to market limitations and uncertainties, the government's attempts to encourage commercialization of the Landsat system have met with much less success than its previous efforts to foster the development of the communications satellite industry. In negotiations with industry, NASA and the Department of Commerce have had difficulty transferring responsibility for operation of Landsat to the private sector, even with the offer of an outright subsidy of $250 million. Projections of potential commercial sales of Landsat data range from $14 million to $75 million a year, at best one to two percent of the revenue levels generated by satellite telecommunications.

Processing Dollars in Space

Over the past five or ten years, the field of space commercialization that has generated the most optimistic projections of revenue and profit has been materials processing in space (MPS). Experiments conducted in 1971 aboard Apollo 14 demonstrated that certain substances can be formed, combined, or separated in the zero gravity environment of space in ways that would be difficult or impossible on Earth. The possibility of creating unique and valuable products in space gave NASA an additional rationale for development of the Space Shuttle, helping to fuel great expectations for the future of space processing. Estimates of the amount of commercial revenue that this technology will generate by the end of the century have ranged as high as fifty billion dollars per year, and these predictions have helped motivate hundreds of entrepreneurs and investors to seek space processing opportunities. NASA has helped stir up this enthusiasm with aggressive programs that support companies investing in commercial space processing applications, and by giving MPS high visibility as an area of interest for the agency. In its 1981-1985 Five Year Program Plan, NASA hailed MPS as one of the areas of space applications holding the greatest "promise for immediate or potential benefits to humanity." More recently, NASA has used commercial space processing as a primary justification for development of its manned Space Station.

But despite claims that materials processing will become the next great growth industry in space, there are grounds for substantial wariness on the part of would-be MPS entrepreneurs. In the decade and a half following America's first MPS experiments on Apollo 14, several hundred million dollars were invested by government, academia, and industry in materials processing research and development. But over this period, the only space-manufactured products to be sold commercially were a few thousand dollars worth of "monodisperse latex spheres," microscopic particles used for calibration of delicate measuring instruments. These polystyrene droplets, each about 1/2500th of an inch in diameter, were produced during a 1984 Space Shuttle mission by a start-up firm, Particle Technology, Inc. in cooperation with NASA. Experiments have shown that many other kinds of materials can also be created in the weightless environment of space more effectively than on Earth, but the basic economics of materials processing in space have thus far limited the number of successful space processing firms to a field of one.

The key obstacle to commercialization of MPS technology is high development costs, which are unavoidable because there is no way to inexpensively simulate the zero gravity environment of space on Earth. NASA's only ground facilities for simulating the effects of weightlessness on materials are drop tubes and drop towers, which enable a small particle to be observed in a state of free-fall for only two or three seconds. An experiment that can be taken to the air without much difficulty can be flown on a parabolic arc in one of NASA's KC-135 aircraft, a maneuver that can provide up to thirty seconds of weightlessness. The only other alternative to putting an experiment into orbit is to launch it on a Space Processing Applications Rocket (SPAR), a small suborbital sounding rocket that can provide about five minutes of simulated microgravity. Facilities such as the KC-135 and SPAR rockets can be useful for relatively simple demonstrations of basic MPS principles, but have very limited value for commercially oriented research and development, which requires many hours of experimentation with close human interaction.

With no known substitute for extended orbital spaceflight, MPS researchers must utilize the Space Shuttle, a reliance that can be very time-consuming and expensive. It generally takes at least three to five years to design and build even the most rudimentary experimental flight hardware, qualify it for flight on the Space Transportation System, and receive a Shuttle launch opportunity. After all this has been accomplished, a launch delay of only a few days can ruin an experiment, as happened to a batch of material flown on the Shuttle by NASA and McDonnell Douglas in 1984. When experiments finally are flown, they may or may not be successful. If they do succeed, a series of several additional flight experiments may be required before the product or products being developed are ready for commercial sale. Over this period, investment requirements can run into the tens of millions of dollars.

NASA's desire to reduce the costs and risks faced by materials processing entrepreneurs resulted in the establishment of the Joint-Endeavor Program in 1979. Through this program, whose provisions are so unique that its implementation required an act of Congress, NASA can sign a Joint Endeavor Agreement (JEA) with any U.S. company interested in developing a commercial use of space. Once NASA and a company enter into a JEA, the agency can provide its industry partner with a variety of resources that can significantly reduce the research and development investments required to establish a successful commercial space enterprise. NASA cannot grant an outright financial subsidy to a JEA participant, but the JEA provisions that NASA can provide may be worth substantial sums of money. NASA's first JEA partner, McDonnell Douglas, has been given free flight time on the Shuttle that would have cost the company tens of million of dollars if purchased at NASA's regular commercial rates.

The objective of the Joint Endeavor Agreement with McDonnell Douglas is to use government resources to help the company develop a profitable business manufacturing high-value pharmaceuticals in space. In addition to free use of the Space Shuttle, the terms of the JEA commit NASA to provide McDonnell Douglas with technical support services and allow the company to use certain NASA laboratory facilities, also at no cost. NASA must also protect any proprietary data generated by McDonnell Douglas over the course of its JEA, and has promised not to sign Joint Endeavor Agreements with any other companies that would support competition with McDonnell Douglas' project. These NASA concessions represented an unprecedented approach to the development a new space technology and to the encouragement of a commercial space activity.

In exchange for NASA's commitments, McDonnell Douglas agreed to invest substantial amounts of its own capital to develop and build the pharmaceutical production facilities that are flown on the Shuttle. The company also committed itself to making a sincere effort to market its space-manufactured products commercially, and agreed to provide NASA with all data gained from its activities if, for any reason, the firm decides to abandon the venture. If the project is successful, McDonnell Douglas must begin paying regular commercial rates for use of the Shuttle as soon as the company begins marketing its products commercially. Company officials have claimed that by the time McDonnell Douglas earns its first commercial revenues, it will have invested tens of millions of dollars of its own money in this risky enterprise.

McDonnell Douglas has kept the exact identity of the first product it intends to produce in space a trade secret, but has revealed that it will be manufacturing pharmaceuticals through a process known as continuous-flow electrophoresis. This is a technique that utilizes electrical charges to separate a desired product from a substance in which it is contained in small quantities. On Earth, the process of electrophoresis is hampered by the force of gravity, but in the weightless environment of space this problem does not exist. As a result, McDonnell Douglas' process has been shown to be about five hundred times more efficient in space than it would be on Earth, where electrophoresis is such a cumbersome process that its products can cost millions of dollars per pound. The total market value of pharmaceutical products that can be produced in space through electrophoresis is projected to be about six billion dollars per year, which explains why McDonnell Douglas has been willing to make the large investments in time and money that have been required to sustain this project.

It is still too early to predict how successful McDonnell Douglas will be in the pharmaceutical production business, nor can the ultimate value of its Joint Endeavor to the Space Program be accurately gauged. The experiments McDonnell Douglas has flown on the Space Shuttle have been highly successful, according to company and NASA officials, but the firm must still contend with a number of problems. Once it has established a working production facility, McDonnell Douglas must expand its production to commercial quantities, obtain approval to sell its products from the U.S. Food and Drug Administration, and compete for a market share with Earth-based manufacturers, whose processes are rapidly improving and who do not have to face the high costs of space transportation. By signing a JEA with McDonnell Douglas, NASA encouraged the company to face all of these risks and established a precedent for the use of government incentives to help similar materials processing in space ventures.

As a result, NASA has received dozens of Joint Endeavor proposals since its pioneering agreement with McDonnell Douglas was signed in 1980, and through 1985 had successfully negotiated and signed JEA's with ten different companies. NASA has shown that it is willing to sign such agreements with small entrepreneurial firms as well as large established companies, as evidenced by nearly half of its JEA's. The agency's third JEA was signed with Microgravity Research Associates (MRA), a small Florida-based company whose 1983 agreement with NASA is aimed at manufacturing crystals in space, using a process called electroepitaxial crystal growth. Like electrophoresis, this process can be carried out more successfully in weightlessness than on Earth, and could result in the production of extremely valuable materials. The primary product MRA intends to manufacture is gallium arsenide, a semiconductor material that may some day become as important to the electronics industry as silicon, and which could sell for as much as $500,000 per pound.

Another small company that signed a JEA with NASA is Cinema 360 Inc., a non-profit consortium of planetariums whose agreement, also signed in 1983, was to produce a film of Space Shuttle activities. This Joint Endeavor is significant in that it set a precedent for application of such agreements to activities other than materials processing in space. The Joint Endeavor program was originally established with the sole intent of promoting the development of commercial materials processing in space, but NASA's agreement with Cinema 360 demonstrated the agency's willingness to negotiate and implement other types of Joint Endeavors. For companies that are not ready to make the investments required for a Joint Endeavor Agreement, NASA can implement Technical Exchange Agreements (TEA's), which allow commercial scientists to work with NASA investigators in areas of applied research. Once a technical breakthrough is made under a TEA, the agreement can be converted into a JEA.

Even with the NASA support that is available, most space processing enterprises are likely to require a multimillion dollar investment before production of a commercial product can begin. Once full scale production is achieved, ongoing operations costs may represent an even greater financial problem for a prospective

MPS investor. Even at subsidized Shuttle prices, a commercial MPS enterprise supported by the STS would have to yield a product worth $10,000 to $100,000 per pound just to recover operations costs. When NASA's Space Station becomes available, the investment required to develop a space product may be reduced, but operating costs are not likely to be affected significantly. As a result, the number and variety of products that can be manufactured in space profitably is likely to remain very limited for the remainder of this century, or at least until lower cost space transportation systems become available.

According to NASA space processing literature, space has potential benefits for manufacturing four types of materials: metals, glasses, and ceramics; fluids and chemicals; electronic materials; and biological materials. Research in all four areas may lead to improvements in Earth processing techniques, but it is unlikely that products in all four categories will be profitably manufactured in space. If scrap iron could be converted into gold through some miraculous space process, it would be done at a financial loss, because gold is currently worth only $4,000 to $5,000 per pound, well below the $10,000 to $100,000 required for MPS profitability. Only electronic materials and biological materials show any real promise for yielding products whose values are in this lofty range, and even here, the numbers are far from overwhelming. Gallium arsenide remains the most highly touted MPS candidate among electronics materials, but hopes of selling this substance for $500,000 per pound are contingent on production of a super-pure, ultra-high performance variety. Lower quality samples of this material currently produced on Earth sell for only a small fraction of this amount.

McDonnell Douglas' area of interest, biological materials, seems to be the only one of NASA's four categories that holds any realistic hope of near-term profitability for MPS entrepreneurs. McDonnell Douglas has identified approximately twenty different pharmaceutical products that it might ultimately manufacture in space, including interferon, skin-growth agents, and insulin-producing cells. Pharmaceuticals are particularly attractive candidates for space processing because they have extraordinarily high values per unit weight. One gram of pure interferon, for example, could provide ten thousand doses, for potential uses in treating ailments ranging from the common cold to cancer. Even at a relatively low price of ten dollars per dose, this translates into a value of *forty-five million* dollars per pound of pure interferon. Another pharmaceutical, synthetic calcitonin, is valued at over one hundred million dollars per pound.

Industry enthusiasts frequently cite the McDonnell Douglas enterprise to lend credence to their projections of tens of billions of dollars in annual space processing revenue. But McDonnell Douglas remains the only company with a reasonable chance of generating large MPS sales in the near future. During NASA's campaign to sell the Space Station program to President Reagan and Congress, the McDonnell Douglas venture was also cited as a great example of the type of activity the Space Station would support. Less than a year after Reagan made his commitment to the Space Station, however, McDonnell Douglas announced that it would not need the Space Station for production of its primary product; it could instead be produced in sufficient quantities on the Space Shuttle. At about the same time, Johnson & Johnson, which had been McDonnell Douglas' partner since the beginning of its project, dropped out of the venture to devote greater attention to ground-based pharmaceutical production opportunities.

If space processing is to follow satellite communications as the next great wave of commercial activity in space, a greater number and larger variety of profitable extraterrestrial processes and products will have to be discovered. This will require a substantial increase in NASA's support for MPS research. In its 1981-1985 Program Plan, NASA projected that annual funding for its Materials Processing in Space Program would rise to over $100 million by 1984. Instead, funding for the agency's MPS Division, which NASA recently renamed the Microgravity Science and Applications Division, remained constant over the entire five-year period at about $20 million per year. Even had the $100 million annual budget been attained, this would have represented less than half of what NASA spent on development of satellite communications technology development two decades earlier. While the agency's innovative Joint Endeavor Agreements and other industry support programs have clearly helped stimulate private sector interest in space processing, the low level of funding for these activities does not seem to be consistent with NASA's stated expectation that commercial MPS will be a major extraterrestrial activity during the Space Station era.

No amount of NASA funding, however, will change the basic economics of space processing, unless major expenditures are directed toward development of a truly economical Space Transportation System. Had the Space Shuttle achieved its original goal of delivering payloads to orbit for less than $500 per pound in 1987 dollars, the value required for a product to be produced in space profitably would be less than half what it is now. NASA's recent decision to curtail its use of the STS for routine launches of commercial satellites will not help reduce Shuttle costs, but it may enable the space agency to focus greater effort on developing a more advanced, cost-effective launch vehicle. It may also provide an indirect long term benefit to materials processing in space in by stimulating the development of a commercial space transportation industry, whose innovations might eventually result in greater reductions in launch costs than the government would otherwise have been able to produce.

Commercial Space Transportation

The misfortunes that have befallen the Space Shuttle have focused widespread attention on the need for a U.S. commercial launch vehicle industry to complement NASA's Space Transportation System. The seeds of a commercial space transportation industry, however, were planted long before the tragic loss of Challenger and her crew. The Reagan Administration established a commercial space policy formally advocating the development of a commercial space transportation industry eighteen months prior to the Challenger disaster, and start-up firms have been proposing ambitious private space transportation enterprises since the mid-1970s. By the time Reagan's commercial space policy was established, at least twenty companies were involved in some significant way in the development or provision of space transportation services on a commercial basis.

The interest of so many companies in space transportation is somewhat surprising, because the obstacles to developing a profitable commercial space transportation system are at least as prohibitive as those facing MPS entrepreneurs. Until recently, would-be space transportation entrepreneurs were forced to contend with major political hurdles as well as massive up-front costs, since a major NASA objective until March 1986 was to capture as much of the world space transportation business as possible with its subsidized Space Shuttle. While NASA's new policy favoring the development of a commercial launch industry is a significant change, it remains to be seen whether any companies will profit from commercial space transportation endeavors in the near future. The space agency has not given any indication that it intends to terminate the subsidies it offers Space Shuttle customers, nor has it gracefully accepted the prospect of transferring commercial satellites that were booked on the Space Shuttle to other launch vehicles.

In the long run, the effects of the Challenger accident may help get a competitive space transportation industry started, but in the short term, the loss of Challenger will have a devastating effect on the commercial space transportation industry. The only commercial space transportation system to earn a profit to date has been the McDonnell Douglas Payload Assist Module (PAM), which flies in the Space Shuttle. The delay in the Shuttle launch schedule and the possible loss of Shuttle customers resulting from the Challenger tragedy may hurt McDonnell Douglas' sales of its popular PAM-D and PAM-D II upper stages significantly. An even bigger loser from the Challenger disaster may be Orbital Sciences Corporation, which had sold only one of its commercial Shuttle-launched Transfer Orbit Stages at the time of the accident. The TOS can be flown on Martin Marietta's Titan launch vehicle, but was designed with the expectation that it would be used primarily as a Shuttle upper stage. If NASA's new space transportation policies portend a reduction in the Shuttle's use for so-called "routine" satellite launches, then the TOS may become a fifty million dollar piece of hardware with no way of getting into space.

The collapse of Orbital Sciences Corporation (OSC) would deal a particularly crushing blow to the commercial space transportation industry, and to space commercialization in general. OSC was founded in 1982 by a trio of youthful Harvard Business School graduates wanting to cash in on the lucrative market for launch services to geosynchronous orbit. When these entrepreneurs discovered that NASA intended to let out a contract to design and develop the Transfer Orbit Stage (TOS), a new upper stage booster for the Space Shuttle, they persuaded NASA Administrator James Beggs to give OSC a chance to prove it could develop the TOS commercially. At OSC's request, Beggs signed a Memorandum of Understanding (MOU) in January 1983 that gave the fledgling firm thirty days to demonstrate that it could put together the technical and financial resources needed to develop the TOS commercially.

This MOU had two important effects: it delayed NASA's procurement of a TOS contract, and it gave OSC an official document that the company could present to prospective investors as a NASA "seal of approval." Armed with the MOU, the young entrepreneurs obtained the agreement of Martin Marietta to develop the TOS under contract to OSC, retained the services of Rothschild, Inc. and Shearson/American Express, Inc., and raised over $50 million in venture capital to finance development of the TOS. Impressed with OSC's resolve and eager to demonstrate its support for space commercialization, NASA scuttled its TOS plans and set up an office at the Marshall Space Flight Center to assist OSC in its project.

The following year, OSC announced that it had signed a second MOU with NASA, for commercial development of the Apogee and Maneuvering Stage (AMS), a liquid-fueled vehicle that can be operated in conjunction with the solid-fueled TOS. By this time, OSC had lured away from NASA and hired some of the space agency's top engineers and managers, including the Director of NASA's Shuttle Upper Stage Division. Full page color advertisements for the TOS and AMS began to appear in industry trade magazines. In the span of a little over a year, OSC had parlayed a single MOU with NASA into a multimillion dollar business, involving some of the nation's largest and most powerful technological and financial corporations. Perhaps even more significantly, OSC had persuaded NASA to relinquish most of its control over a project that the agency had intended to develop through conventional contractual arrangements. With all of the publicity and high hopes generated by this enterprise and the tens of millions of dollars in venture capital at stake, the failure of OSC's project could have a very discouraging effect on potential investors in future commercial space projects.

The only sure winner amidst the current realignment of the

space transportation industry is Arianespace. When it became apparent that the loss of Challenger would keep NASA out of the transportation business for an extended period, the French quasi-private consortium immediately announced a thirty percent price increase for its Ariane expendable launch vehicle (ELV). At the time, Arianespace was the only organization in the world prepared to offer commercial launch services, largely because American ELV manufacturers had been effectively put out of business by the French government's subsidies to Arianespace and NASA's cutthroat Shuttle pricing policies. General Dynamics, which began actively marketing its Atlas/Centaur launch vehicle commercially in early 1983, had discovered that it could not sell any vehicles because its price was thirty to fifty percent higher than the subsidized Ariane and Shuttle prices. Companies attempting to market McDonnell Douglas' Delta ELV and Martin Marietta's Titan launch vehicle encountered similar difficulties, and as a result, production lines for these vehicles were slowed to a virtual standstill.

When the Shuttle accident occurred, it was not possible to rapidly shift the Shuttle's payloads to the Atlas/Centaur, Delta, and Titan, because these vehicles all require about three years for production. Nor were their manufacturers eager to finance the production of vehicles on their own. For decades, General Dynamics, McDonnell Douglas, and Martin Marietta had been building these rockets for NASA on a contractual basis, with NASA making progress payments to each contractor to cover production costs as they were incurred. When NASA stopped awarding these contracts, the ELV manufacturers could have invested their own money in building additional rockets, but this would have represented a radical departure from the normal business practices of these companies. It also would have been risky, since no customers were willing to commit to buying these ELV's as long as much lower Shuttle and Ariane prices were available.

The Department of Defense, which had foreseen possible Shuttle limitations in the early 1980s, had contracted with Martin Marietta and General Dynamics to build ten Titan/Centaur "CELV's" (complementary expendable launch vehicles) in 1985, but these vehicles were earmarked for specific military missions in the early 1990s, and therefore were not available to ease the launch vehicle shortage caused by the grounding of the Shuttle fleet in 1986. In its determination to use the Shuttle for every possible mission, NASA had even opposed the purchase of these CELV's by the military, and had to be overruled by the President before the DOD was allowed to finalize its purchase.

NASA was able to monopolize the American space transportation industry without significant industry resistance because the established ELV manufacturers remained dependent on NASA for other space business, and did not wish to imperil this business by challenging the policies of the space agency. In fact, all three major ELV producers decided to make the best of the situation by developing new transportation systems to be flown out of the Shuttle: McDonnell Douglas' PAM, Martin Marietta's TOS, and General Dynamics' Shuttle/Centaur. The only American companies to voice significant challenges to NASA's transportation policies were the new start-up ELV firms that began appearing in the early 1980s. The first of these companies was Houston-based Space Services, Inc., which symbolically ushered in the commercial space transportation era on September 9, 1982 by launching its privately-assembled Conestoga I rocket from Matagorda Island off the coast of Texas. After a ten minute flight to an altitude of 196 miles, the rocket and its mock payload of forty gallons of water splashed down in the Gulf of Mexico 321 miles from its launch site.

On August 3, 1984, Space Services' feat was duplicated to an extent by Starstruck, Inc. of Redwood City, California, which achieved an unusual water launch with its Dolphin rocket in the Pacific Ocean just west of San Diego. The rudimentary Dolphin reached an altitude of only 3,000 feet before veering off course, forcing Starstruck's engineers to end its flight. This modest accomplishment apparently did not impress the investors who had sunk fifteen to twenty million dollars in the Dolphin's development since 1981; about two months after the test launch the company collapsed and sold its remaining rocket hardware to a space museum. Space Services has managed to avoid Starstruck's fate, but is far from mounting a serious threat to Arianespace. When last heard from, the company announced that its first commercial venture would be to launch the cremated remains of thousands of people into space. Despite the modesty of their achievements, Starstruck and Space Services remain the only two companies to accomplish flight tests, out of the several small companies trying to enter the launch services business. The other companies are all at earlier stages in the development of their systems, every one of which will require an investment in the tens of millions of dollars to become operational.

One of the major objectives of President Reagan's 1984 commercial space policy was to help both large and small companies develop private launch vehicles. A Commercial Space Transportation office was formed within the Department of Transportation to reduce the obstacles to commercialization of such systems. The new office promptly claimed credit for obtaining the bureaucratic approvals necessary for Starstruck's test flight, but has been largely ineffective in removing the two greatest barriers to launch vehicle commercialization — high development costs and NASA's Shuttle pricing policy. Entrepreneurs are attracted to the space transportation industry by the potential for collecting the enormous fares that customers are willing to pay, but they quickly learn that the cost of developing and testing space vehicles is also great. The few aerospace companies whose systems are fully developed have realized that even the efficiency of private operation cannot overcome the disadvantages created when governments subsidize competing vehicles.

With subsidized competition suspended at least temporarily by the grounding of the Shuttle fleet, the nation's established ELV manufacturers have gradually moved into the unfamiliar territory of commercial marketing. In mid-1987, General Dynamics elected to build eighteen Atlas/Centaur vehicles with company funds, initiating the largest commercial space venture in history. Once General Dynamics determined that it would not have to compete with a subsidized national transportation system, the risks associated with commercializing its proven launch vehicle became acceptable.

But for companies to successfully face the risks of developing new space transportation systems commercially, the government may need to go further than simply repealing its competition-suppressing policies. NASA may need to actively help private companies develop key transportation technologies, just as it has helped advance the technologies of satellite communications, remote sensing, and materials processing in space. The amount of money NASA spends fostering commercial activity in these other areas, if invested in transportation technologies, would be sufficient for the development of new vehicle systems, launch facilities, and test equipment that could be used by private transportation companies to reduce their investment requirements. To achieve a proper balance, the space agency should devote some of its resources to developing the transportation systems of the future, such as the National Aerospace Plane, while simultaneously helping private companies develop near-term commercial applications for state-of-the-art technologies. The future of space transportation, and of space commercialization in general, will depend on how well NASA adapts its policies to the changing needs of space age entrepreneurs.

6

New Directions for NASA

The best single cure for the problems that plague America's Space Program is for private industry to assume a greater share of the risks and rewards of space development. But if private industry is to have broader responsibilities for the operation of space systems, NASA will have to offer greater encouragement to the private sector. To effectively promote "privatization," the space agency might even need to make fundamental changes in the way it normally conducts its business. NASA has established a number of innovative programs to help encourage private firms to invest in space development, but the agency has shown a reluctance to support commercialization when control over a major program is at stake. As an example, NASA has maintained total control over Space Shuttle operations and is using taxpayer money to fund the production of a new Orbiter to replace *Challenger,* despite the fact that the agency received several proposals from private companies interested in building additional Space Shuttle Orbiters or operating the Shuttle for profit. The interest of private companies in operating the Shuttle, and the failure of the space agency to achieve economical Shuttle operations, raise the question of whether NASA is the ideal organization for carrying out ongoing operations. It also poses the question of whether NASA's approach to developing future systems, such as the Space Station, should be changed to give the private sector a greater role in their development and operation.

Should NASA Continue to Operate the Space Shuttle?

Development of the Space Shuttle was a task so complex and costly that it almost certainly could not have been achieved as a totally private enterprise. Soon after the Space Transportation System became operational, however, entrepreneurs began expressing interest in taking over certain operational responsibilities. The first proposal for private operation of the Space Shuttle was presented to NASA by a start-up firm calling itself *Space Transportation Company* in early 1982, before the Shuttle had even completed its four early test flights. Space Transportation Company (STC) was founded by space entrepreneur Klaus Heiss, whose economic analyses in the early 1970s had helped NASA sell the Shuttle program to Congress. Through STC, Heiss was able to enlist the support of such high-powered financial backers as the Prudential Insurance Company, and developed an ambitious plan to fund the purchase of a fifth Shuttle Orbiter with private investment capital. According to STC's plan, this privately-financed Orbiter would essentially be turned over to NASA to augment the agency's fleet of four taxpayer-funded Orbiters. In exchange for its financing of the fifth vehicle, STC proposed that NASA grant the company marketing rights to ten Shuttle launches per year.

By "marketing rights," STC meant that it would assume responsibility for finding enough customers to utilize the company's ten annual Shuttle launches. STC also proposed to assume the obligation of collecting Shuttle use fees from these customers, which the company pledged would not exceed the price that would otherwise be charged by NASA. But when NASA evaluated STC's proposal, the agency determined that STC's participation would not significantly reduce the overall cost of operating the Shuttle fleet. STC would be able to recover its billion dollar-plus investment only by obtaining from NASA an additional premium for flights of the fifth Orbiter. STC proposed to accomplish this by reimbursing NASA for only the *variable* costs of operating the fifth Orbiter. By so doing, STC would essentially pocket the equivalent of NASA's fixed operations costs, an estimated thirty million dollars, for each of these ten flights per year. Not surprisingly, NASA was unable to identify any significant benefit that STC's plan would provide to the agency or America's taxpayers, and allowed the first proposal for private "operation" of the Shuttle to die quietly.

However, NASA's rejection of STC's bid did not dampen private industry's interest in becoming involved in commercial Shuttle production and operations. In March 1984, Cyprus Corporation, a Pittsburgh-based investment firm, initiated negotiations with NASA for a private purchase of a fifth Orbiter, with the intent of selling insured launch services to commercial customers. Cyprus, which has since changed its name to Astrotech International, is headed by Willard F. Rockwell, Jr., whose family founded Rockwell, International, the giant conglomerate that builds the Shuttle Orbiters under contract to the space agency. Astrotech's Space Shuttle of America Corporation, a subsidiary formed exclusively for the purpose of implementing this transaction, reiterated its interest in the multibillion dollar purchase of an additional Orbiter after the loss of Challenger in January 1986. Despite this offer, NASA lobbied intensively for government appropriation of sufficient funds to build a replacement Orbiter, and during the summer of 1986 the Reagan Administration responded by approving this request.

Appropriation of government funds to replace Challenger dashed any lingering hopes that a privately-financed Orbiter would be added to the Shuttle fleet. But even if the government continues to finance the production of all Shuttle Orbiters, America's taxpayers could accrue significant benefits if day-to-day operation of the Orbiter fleet could be transferred to a private company. The STC and Astrotech examples show that American entrepreneurs are willing to invest in the costliest and riskiest areas of space transportation, with reputable businessmen giving serious consideration to multibillion dollar investments. These space industrialists and their backers agree that NASA should not be in the business of operating a "routine" transportation system. While it is not clear that operation of the Shuttle is in any sense routine, particularly in view of the Challenger accident, the argument that NASA should not have total responsibility for the indefinite operation of a major space system certainly has some merit.

When NASA was founded in 1958, the agency was envisioned as having three basic responsibilities: scientific exploration, development of new technologies, and establishment of national space policies. For the first twenty-five years of NASA's existence, it was relatively easy to determine which space activities were the rightful responsibility of private industry. By 1961, for example, it was generally agreed that manufacture and operation of communications satellites were in this realm, so responsibility for these activities was rapidly transferred to the private sector. Conversely, NASA's major programs of the 1960s and 1970s — Mercury, Gemini, Apollo, and Skylab — were indisputably dedicated to achievement of broad scientific, technological, and political objectives, and held little attraction as private initiatives.

Development of the Shuttle was also consistent with NASA's charter as a research and development oriented entity, because the effort relied on development of new technologies and added a new dimension to mankind's capabilities for space exploration and utilization. The Shuttle development effort also resembled its predecessors among NASA's major programs in that it had a specific goal that would be accomplished over a well defined time period. It was assumed that once the goal of construction of the Orbiter fleet was achieved, government funding for the program would no longer be needed. Space Shuttle customers would bear the full cost of operating the nation's Space Transportation System, and NASA would dedicate its resources to other challenges, such as development of the Space Station.

The unexpectedly high cost of operating the Shuttle, however, made this impossible. Had the Shuttle met its economic goals, the space agency would have been stampeded by private entrepreneurs eager to operate the system for profit. Responsibility for operating the Shuttle could have been transferred to the private sector, leaving NASA free to develop new programs. But paying for the high cost of operating the Shuttle required a continued government subsidy, helping to solidify NASA's control over the program. Realizing that its budget requests for Shuttle operations were sapping money away from other programs, including its coveted Space Station project, NASA officials became even more consumed by their desire to reduce STS costs by increasing the Shuttle flight rate. In their efforts to achieve higher flight rates, NASA officials steered the space agency still further away from its intended role as a research and development organization.

Rather than orient the Shuttle toward those applications that were most consistent with the science and technology-related tasks that the agency was established to pursue, NASA attempted to fly as many Shuttle missions as possible by aggressively marketing the system to potential commercial customers. To attract these customers, the agency recruited a cadre of salespeople, whose time was spent developing slogans and pricing policies, rather than new and better ways of gaining access to space resources. To make the Shuttle economically attractive to its customers, NASA devoted nearly a third of its entire budget to the provision of STS price subsidies. If not for the slowdown caused by the Challenger accident, the agency would have spent more money subsidizing its STS customers during the 1980s and 1990s than it would have devoted to development of its Space Station.

But all of this changed in the aftermath of the Challenger accident. Six weeks after the tragedy, NASA's Acting Administrator William R. Graham announced that the space agency would alter its longstanding insistence on flying as many payloads as possible on the Space Shuttle. In a memorandum to the director of the Shuttle program, Graham formally acknowledged NASA's reversal by expressing support for a "mixed fleet of Orbiters and E.L.V.'s" and by endorsing "the development of a viable, competitive, domestic commercial capability" for providing launch services.

NASA was forced to accept these new policies as a result of the delays, reduced Shuttle launch capacity, and general safety concerns that were caused by the Challenger accident. Reliance on a government-operated Space Shuttle as the nation's only means of access to space was now clearly shown to be inconsistent with the public's interest. In view of these new circumstances, it was generally agreed that the most logical payloads to be removed from the Shuttle's manifest would be those commercial spacecraft that could be orbited using unmanned expendable rockets. Many in the industry argued further that the Shuttle should henceforth be used only to launch payloads that absolutely require manned interaction. Proponents of this approach continue to argue that a new generation of unmanned vehicles should be developed to fly payloads that are too large to be flown on current ELV's.

It is unfortunate that a tragedy of the magnitude of the Challenger accident was required before NASA's excursions into the commercial marketplace were brought to a halt. NASA was not created to compete with private companies in any kind of on-going space business. By attempting to do so, NASA turned the Shuttle into a confusing hybrid of a national research and development resource and a commercial transportation system. If the Shuttle re-emerges as a commercially attractive vehicle, then it should be operated by private companies for private companies.

Regrettably, the prospects of operating the Shuttle for profit now are even slimmer than in the pre-Challenger years. The customers that were once considered most likely to make the Shuttle cost effective — commercial satellite owners with mature, profitable businesses — will be those most likely to turn to ELV's to meet their launch needs. The Department of Defense also seems determined to reduce its reliance on the Shuttle by using expendable rockets to deliver millitary payloads to orbit. With such a large loss of commercial and military customers, it is unlikely that the Shuttle flight rate will ever go much higher than ten missions per year. At such a low rate, the cost of each mission will probably remain in the vicinity of $250 to $300 million. Transferring responsibility for Shuttle operations to a private company would make sense only if substantial reductions in these costs could be achieved as a result. Since a private Shuttle operator would have to earn a profit to stay in business, actual operating costs would need to be reduced by at least twenty to thirty percent for Shuttle privatization to significantly reduce the cost of STS operations to America's taxpayers.

Whether private operation of the Shuttle could result in such cost reductions is questionable, but is nonetheless a question that deserves a thoughtful answer. The cost of constructing key Shuttle hardware elements, such as the external fuel tanks and solid rocket boosters, could conceivably be reduced if the method of procuring these items were modified. A private Shuttle operator could, for example, provide the manufacturers of the external fuel tank and solid rocket boosters with greater incentives to reduce production costs, perhaps by offering a share of STS operating profits as an inducement. A private operator might also be able to reduce the size of the work force NASA currently employs to operate the Shuttle. Military officials have claimed that they could reduce the size of this "standing army" by a factor of five. NASA officials have opposed the transfer of Shuttle operations to the private sector largely for this very reason — it would eliminate the need for thousands of agency jobs.

There is only one way to reliably evaluate the feasibility of operating the Shuttle privately. To accomplish this, the U.S. Government would have to formally offer to turn the entire operation over to industry and assume the role of a Space Shuttle customer. This could be achieved if the government offered the use of its Orbiters and Shuttle launch facilities to prospective industry Shuttle operators as government-furnished equipment, and declared its intent to lease a specified number of Shuttle launches from the lowest bidder. This would motivate a number of qualified companies to perform detailed studies of how the Shuttle could be operated privately, with emphasis on identifying methods to achieve the lowest possible operating cost. The government would almost certainly receive a strong response to such a solicitation, because several companies currently involved in operating the Shuttle as NASA contractors would have their businesses at stake. Under this new lease-back arrangement, the private Shuttle operator would have responsibility for subcontracting for Shuttle hardware and services in a manner of its own choice, and would market and sell Shuttle launch services to commercial and foreign customers at negotiated terms.

Under such a scenario for privatization of Shuttle operations, the government could reserve its right to reject all bids, if no company offered a technically sound proposal with a lease-back price that would reduce taxpayer costs. This would indicate that NASA's current Shuttle operating system is as efficient as possible, given present market conditions and the state of technology. The cost of operating the Shuttle could then be considered merely one of many expenses incurred by the U.S. Government to improve the security and quality of life of the American people. If continued NASA operation of the Shuttle proved to be detrimental to the agency's ability to effectively develop new technologies and space systems, then responsibility for this task could be given to another government organization, such as the Department of Transportation or the Federal Aviation Administration. Alternatively, an entirely new agency could be formed to operate the Shuttle, perhaps by splitting NASA into two separate agencies, one of which would retain responsibility for space research and development.

Whether private operation of the Shuttle could succeed remains a mystery, primarily because the system was not designed to be operated privately. From the earliest design stages of the Shuttle program onward, NASA assumed that it would manage Shuttle operations, relying on contracted industry support for specified tasks. The agency did not offer Shuttle development contractors any significant incentives to make the Shuttle cost-effective to operate, nor did NASA encourage its industry Shuttle team to consider the possibility of private operation. To further thwart any hopes that the Shuttle would be a commercially attractive system, NASA was forced by budget constraints to make design compromises that effectively assured that the Shuttle would be very expensive to operate. As a result, the true cost of delivering payloads to orbit on the Shuttle is no less than performing the same task with the ELV's the Shuttle was designed to replace.

To ensure that the Shuttle is operated in the manner that is most efficient and consistent with America's long range goals in space, government and industry should work together to identify and evaluate all possible options for private operation of the Shuttle, perhaps through a trial solicitation such as the one just described. But it is even more imperative that NASA take immediate steps to assure that private industry is involved in the operation of *future* space systems in the most productive manner possible. Had the space agency considered privatization of Shuttle operations to be a high priority during the 1970s, long-range plans could have been established and implemented to make such a transition work smoothly and effectively. While it may be too late to make the Shuttle an attractive system for private industry to operate, NASA can apply the lessons learned from its Shuttle experiences to its planning for future programs, beginning with its next major effort, the Space Station.

The Space Station: A New Opportunity for Innovation

The Space Station offers NASA a second chance to get private industry involved in the development and operation of a major space system. By the early twenty-first century, the Space Station will serve as a focal point for virtually all human activities in space, many of which will represent commercially attractive business opportunities. For this potential to be realized, the mistakes that have reduced the Space Shuttle's usefulness to private industry must not be repeated. Unfortunately, NASA officials are planning to develop and operate the Space Station in essentially the same way that the Shuttle program was implemented. NASA has tried to give the Space Station program a different image through its "design-to-cost" approach, which will force the size and functions of the Space Station to be determined largely by pre-established cost limitations. But in view of the way budget constraints impaired the Shuttle design, design-to-cost may turn out to be a risky way to set Space Station priorities.

Moreover, design-to-cost will not have any effect on the way the Space Station is operated. If NASA follows its current plans, the private sector will play the same subservient, low risk role in Space Station operations that it has during the Shuttle program. During the development phase of the program, design-to-cost will provide each of NASA's Space Station contractors with incentives to control up-front development costs. These companies might even be offered financial incentives to help NASA meet its design-to-cost targets. But controlling Space Station operations costs is a much lower priority goal for NASA than limiting up-front program costs. The space agency appears determined to reserve for itself the responsibility for operating the entire Space Station, even at the expense of rejecting private sector initiatives that cost reduce the cost of establishing a manned presence in space. According to current agency plans, a number of contractors will develop and build Space Station elements, and the obligations of these companies will be fulfilled once this hardware is accepted by NASA for integration into the Space Station system. In fact, until recently NASA was actually attempting to increase its role on the Space Station program by acting as the "systems integrator," a responsibility often delegated to industry. It wasn't until the aftermath of the Challenger accident that NASA officials conceded that industry should be given this role.

NASA has estimated that it will cost approximately half a billion dollars a year to operate its initial Space Station. This figure represents some creative accounting, because the four Shuttle flights required each year for Space Station resupply and crew rotation may alone cost twice this amount. What half a billion dollars per year does represent is the maximum amount of money NASA officials believe they will be able to devote to Space Station operations, and still have enough funds left over to develop new programs. To provide assurance that the cost of operating the Space Station will not further cripple NASA's research and development abilities, the space agency is apparently trying to convince itself and everyone else that the station will be inexpensive for the Government to operate. NASA appears uninterested in the alternative approach: to establish a facility that can be operated privately, transferring operational responsibilities to the private sector.

If the Space Station is developed as currently planned, responsibility for part or all of its operations could possibly be transferred to the private sector at some date after completion of its construction. However, this would be just as difficult as transferring control over the Space Shuttle would be today. Privatizing Space Station operations would almost certainly be more efficient if NASA established private station operations as a goal during the development phase of the program. Space Station contractors and other companies could then get an early start at developing relevant technical capabilities and business strategies, reducing the time required for such a transition to be implemented. The development of a large government bureaucracy to support Space Station operations could perhaps be avoided, saving taxpayer money and reducing economic and political pressure to keep such activities in the public sector.

An early statement of the U.S. Government's intent to privatize Space Station operations could have an even greater benefit, by enabling cost-saving changes in procurement during the Space Station development and production phases. Award of contracts for production of Space Station hardware could be made conditional upon contractor agreement to assume certain risks and responsibilities associated with operation of the station. NASA officials have argued that a government commitment to finance the entire Space Station must be made before companies will invest their own resources in related commercial projects. But this claim is questionable, as evidenced by several industry initiatives to develop Space Station-related elements commercially.

In August 1983, NASA signed a Joint Endeavor Agreement with Fairchild Space and Communications Company for development of a free-flying platform called the "Leasecraft," which could be used to process larger quantities of materials, for longer periods of time, than the Space Shuttle. Fairchild had hoped that McDonnell Douglas' pharmaceuticals would represent a large share of its early Leasecraft business, and designed a platform that would meet McDonnell Douglas' materials processing requirements. As the McDonnell Douglas project progressed, however, it became evident that the company's pharmaceutical production would not require the capabilities of the Leasecraft, and with no other customers in sight, Fairchild's venture collapsed. NASA had helped ensure the Leasecraft's fate by continuing, as part of its Space Station efforts, to design free-flying platforms with many features identical to those of the Leasecraft. To continue the Leasceraft program without McDonnell Douglas as a firm customer, Fairchild would have risked competing for an uncertain amount of business with very similar, NASA-operated, taxpayer-subsidized Space Station platforms.

NASA's approach to developing the Space Station is also hindering the efforts of Space Industries, Inc. (SII) to develop a man-tended materials processing experimentation and production module. SII has raised several million investment dollars since its formation in 1982 and has prepared a detailed design for an "Industrial Space Facility" (ISF), a configuration of Shuttle-launched and supported modules that are almost identical to those NASA is designing for its Space Station. After several years of pressure, NASA finally signed an MOU pledging modest support for SII in August 1985, but the agency has not made any plans to integrate SII's activities into its Space Station program. NASA's lack of support contributed to difficulties that SII encountered in its attempts to find financing and technical support for the ISF, which, like the Leasecraft, could end up competing for a limited number of customers with subsidized elements of a NASA Space Station. SII eventually teamed up with 3M Corporation and Westinghouse to develop the ISF.

Fairchild and Space Industries are only two of many companies whose activities could be incorporated into the mainstream of NASA's Space Station activities. NASA is continuing to develop the Orbital Maneuvering Vehicle with taxpayer money, despite the fact that the basic design of the OMV is very similar to that of the Apogee and Maneuvering Stage that Orbital Sciences Corporation is developing. The two systems are almost exactly the same size, utilize the same type of propellants (the hydrazine mixture used by the Shuttle's Orbital Maneuvering System engines), and can even perform the same types of missions. The most significant difference between NASA's Orbital Maneuvering Vehicle and OSC's stage is that the AMS is being developed without taxpayer subsidy. If NASA were to rely on the AMS instead of developing the OMV, Orbital Sciences would have the opportunity to serve the future market for satellite servicing without concern over competition from a NASA-subsidized stage.

The commercial value of satellite servicing and other types of Space Station operations might eventually dwarf the revenue from materials processing, opening up a wide variety of opportunities for entrepreneurs interested in on-orbit operations. The Space Station and space platforms will be used by the mid-1990s for the transportation and servicing of all types of spacecraft, ranging from common communications satellites to the most complex scientific observatories. The orbital transfer vehicle will complement the OMV, enabling these spacecraft to be moved from one orbit to another or repaired in space with minimal reliance on costly deliveries of equipment and supplies from Earth. By the early twenty-first century, such transportation and servicing operations will probably be carried out frequently and routinely. NASA's recent transportation architecture studies concluded that by the Space Station time frame, fifty to one hundred such activities might need to be carried out each year. The revenue potential from space based transportation and servicing is in the billions of dollars per year.

Even operation of the Space Station itself could become a profitable commercial activity. A "space hotel company," for example, might operate and resupply the station's habitat modules, recovering its investment by charging "rent" to visiting astronauts on a weekly or monthly basis. A "space power company" could become the first space based utility, operating a solar power system and selling electricity to the residents and users of the Space Station. NASA could turn these operations over to private companies after the Space Station is built, or could provide incentives to encourage companies to invest their own resources in developing these systems. NASA could do this by establishing a "market guarantee," whereby the agency would promise to use these facilities at some minimum rate for a mutually acceptable user fee. With private capital used to develop certain Space Station elements and services, NASA would be able to use a greater share of its resources for developing scientific and other uses for the station. This technique was used by NASA to develop the Tracking and Data Relay Satellite System (TDRSS), which was financed privately after NASA agreed to lease the system for a period of ten years. This arrangement may in fact have saved the first TDRSS satellite from becoming lost irretrievably in space in 1983. After the IUS booster that was supposed to deliver the TDRSS spacecraft to orbit failed, the satellite was nudged into its proper orbit by using additional propellants that its builder had provided for possible commercial applications of the satellite.

By using proven incentives such as market guarantees and Joint-Endeavor Agreements, the U.S. Government could increase its reliance on the same entrepreneurial motivation that has helped develop other frontiers, assembling the Space Station through a series of joint ventures. This would not only assure a private sector stake in operating the Space Station efficiently, but would also reduce the costs to the taxpayers of developing the Space Station. The lessons learned by developing new means of government-industry partnership in the Space Station program could then be applied to future NASA endeavors to utilize the resources of space, including development of more advanced space transportation systems, large scale space manufacturing, and establishment of permanent human settlements in space. Greater reliance on the benefits of free enterprise would be consistent with America's proven approach to the development of new frontiers, and would almost certainly help make long term space dreams such as these a reality.

The Right Role for NASA

The ventures by Fairchild, Space Industries, and Orbital Sciences confirm that private industry has developed a genuine interest in commercial projects related to the Space Station. As NASA's new projects become increasingly vulnerable to pressures to reduce federal spending, the space agency may be forced to reevaluate its lukewarm responses to these companies. If NASA so desired, the agency could save millions of dollars by developing a Space Station architecture incorporating privately-designed and existing systems into its basic design. The space agency could formalize the process of acquiring commercially-developed Space Station elements by issuing "Requests for Joint Endeavor Proposals," which would enable companies to bid competitively for the right to negotiate with the government for the provision of specified equipment and services. Through this process, which NASA has employed in the past to stimulate competition, the companies submitting the best designs and most attractive terms could then be selected.

Commercial development of Space Station elements would probably entail far greater costs and risks than any of NASA's Joint Endeavor partners have assumed to date, so these agreements might have to be modified to provide greater incentives. One approach would be for NASA to give an industry partner a "license" to operate a Space Station element for profit for a specified period, perhaps with an additional market guarantee — a commitment on behalf of the government to utilize this facility and pay a rental fee to the industry partner. Under such an arrangement, NASA could maintain an active role by managing the Space Station systems integration contractor. The space agency would also develop through conventional government procurement processes any Space Station elements that were not sufficiently attractive as business opportunities to be developed commercially. NASA has not adopted this sort of strategy, primarily because many agency managers fear it would reduce their control over the design, development, and operation of the overall system. In a 1984 report on the Space Station, the Congressional Office of Technology Assessment criticized NASA for carving out a multi-billion dollar Space Station empire, leaving little room for entrepreneurial investment in development or operation of the station.

A key recommendation of this report was that NASA place less emphasis on performing Space Station tasks that could be accomplished by private companies.

While use of government incentives to develop systems such as the Space Station would entail new risks and potential problems, none of these difficulties seem insurmountable. The most significant concern is that NASA's industry partners, in their efforts to reduce development and operations costs, could produce systems less safe and reliable than those that would be created under the conventional procurement process. To avoid this possibility, it might be necessary for NASA to establish some means of verifying compliance with safety standards without infringing upon the "commercial sovereignty" of the companies involved. NASA's current procedures for assuring the safety of commercial payloads flown on the Space Shuttle could provide a basis for developing the types of future arrangements that would be necessary if Space Station elements were developed commercially. NASA's past experience in flying commercial spacecraft, which rarely cause significant safety problems, suggest that the safety issue should not serve as a deterrent to developing more creative approaches to the procurement of space systems.

A more plausible difficulty is the possibility that a company developing a Space Station element or other important system might discontinue its development of that system or go out of business completely. This has already happened at least twice in the past, with the folding of the Fairchild Joint Endeavor and the collapse of another Joint Endeavor, with San Diego-based GTI, Inc. This latter agreement, NASA's second JEA, was signed in late 1982 with the expectation that GTI would use investment capital to develop and market a small alloy solidification furnace, which would be flown on the Shuttle to investigate the effects of weightlessness on the formation of exotic metals. GTI's goal was to perform dozens of such experiments for a number of customers simultaneously, collecting a fee of several thousand dollars for each alloy sample created. GTI's proposed role as a "middle man" was similar to the role a company would assume as the developer and operator of a Space Station materials processing module, except on a much smaller scale. Unfortunately, GTI's project collapsed when it became apparent that the market for its alloy solidification furnace was nonexistent.

If a major NASA program such as the Space Station depended on the successful completion of commercial or joint ventures, precautions would need to be taken to reduce the likelihood of this type of outcome. This could probably be achieved most effectively through the exercise of greater discretion by NASA. From an economic point of view, both the Fairchild and GTI ventures were questionable from the start, and were approved by NASA basically because the agency had little or nothing to lose if the enterprises failed. If the stakes were higher, NASA's approval of these proposals almost certainly would not have come as easily. To win NASA's support for commercial development or operation of a critical space system, a company would have to demonstrate capability, commitment, and some means of enabling NASA to continue the program if the company voluntarily or involuntarily ended its involvement.

Through its Joint Endeavor Agreements and similar arrangements, NASA has taken steps to adapt itself to the changing market for space systems and services, an environment in which commercial enterprises are playing an increasingly important role. NASA has also shown sensitivity to the needs of its Space Shuttle customers by implementing attractive pricing policies, and to its prospective Space Station customers by making their needs important considerations in the design and development of its next major program. But the agency has demonstrated little interest in helping to commercialize systems, such as the Space Shuttle and expendable launch vehicles, that would diminish NASA's role as the primary provider of transportation services. NASA also appears intent on maintaining its role as financier, designer, developer, and operator of the nation's Space Station. Since NASA's employees do not earn profits from operation of the Shuttle nor from Space Station development, there is no logical reason for NASA to covet the dominating role it has sought on these programs, particularly if this role is pursued at the expense of successful commercial ventures that could reduce costs to taxpayers and users of space systems. In the current reevaluation of NASA's role in shaping the future of America's Space Program, consideration should be given to the proven and potential benefits of commercialization, and to ways of changing NASA's role so that the maximum possible benefits of private enterprise in space are achieved.

The final Atlas/Centaur test flight, in preparation for the vehicle's first operational flight on May 30, 1966 — the Surveyor I mission to the Moon. In the largest commercial space venture in history, General Dynamics decided in 1987 to build eighteen of these vehicles, for sale to government and commercial customers worldwide.

An overhead view of Skylab, America's first space station, as photographed by the final Skylab crew before their return to Earth in February 1974. Evident in this photo is the missing left solar panel, which was lost during launch of the Skylab workshop.

The Mobile Launch Platform at the Kennedy Space Center transporting the Orbiter Columbia, in preparation for the STS-5 mission in 1983. The MLP transports the Space Shuttle from the Vertical Assembly Building (VAB) to the launch tower at the Kennedy Space Center.

Floodlights illuminating the Orbiter Challenger as it sits on the Mobile Launch Platform at night prior to the launch of STS-7 in June 1983. Moving at one mile per hour, the MLP takes eight hours to make the eight mile journey from the VAB to the launch pad.

The seventh launch of the Space Shuttle and the second lift-off of the Orbiter Challenger, on June 18, 1983. One of the five crew members on this mission was Dr. Sally K. Ride, America's first woman in space.

A view of a Space Shuttle launch prominently showing the operation of the powerful solid rocket booster (SRB). The Shuttle's two SRB's provide a total of five million pounds of thrust, propelling the Orbiter and its external fuel tank through the dense lower atmosphere.

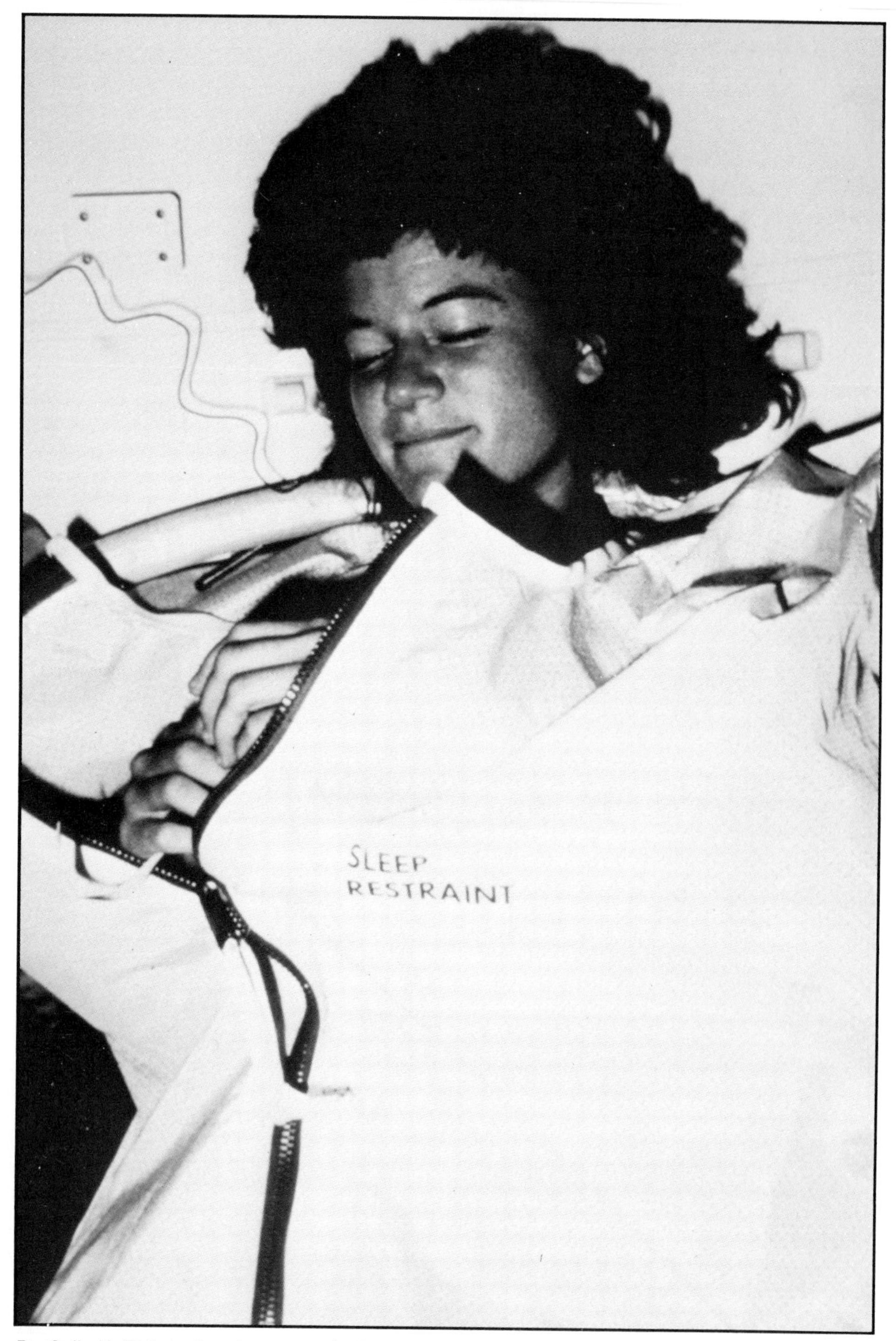

Dr. Sally K. Ride in the sleep restraint bag used by astronauts during Space Shuttle missions. Dr. Ride, who flew on STS-7 in 1983 and STS 41-G in 1984, left NASA in 1987 to work at Stanford University.

Deployment of the Morelos-B communications satellite for Mexico during STS 61-B in November 1985. This was the second flight of the Orbiter Atlantis and the next-to-last successful Shuttle mission prior to the Challenger accident.

Retrieval of the Westar VI communications satellite and its PAM-D upper stage by astronauts Dale A. Gardner (left) and Joseph P. Allen (right) during STS 51-A in November 1984. The satellite had been launched in February 1984 by STS 41-B, but was left in a useless low Earth-orbit when the PAM-D rocket failed to fire.

The Orbiter Challenger as photographed from the unmanned SPAS-01 satellite, which performed various experiments during the seventh Space Shuttle mission in June 1983. After performing these experiments, the SPAS pallet was retrieved using the Shuttle's Remote Manipulator System (RMS) and returned to Earth in the Challenger cargo bay.

The Long Duration Exposure Facility (LDEF) orbiting the Earth in the vicinity of the Orbiter Challenger, which launched the LDEF in April 1984. The unmanned LDEF module houses dozens of automated experiments.

The Spacelab module in the Orbiter cargo bay during processing of the Orbiter for launch. Of the dozens of Spacelab missions originally planned for the 1980s, only three were flown prior to the Challenger accident in 1986.

The four science specialists on NASA's first Spacelab mission, STS-9, are shown viewing a television monitor in the Spacelab module. From left to right are Robert A. R. Parker, Byron K. Lichtenberg, Owen K. Garriott, and Ulf Merbold.

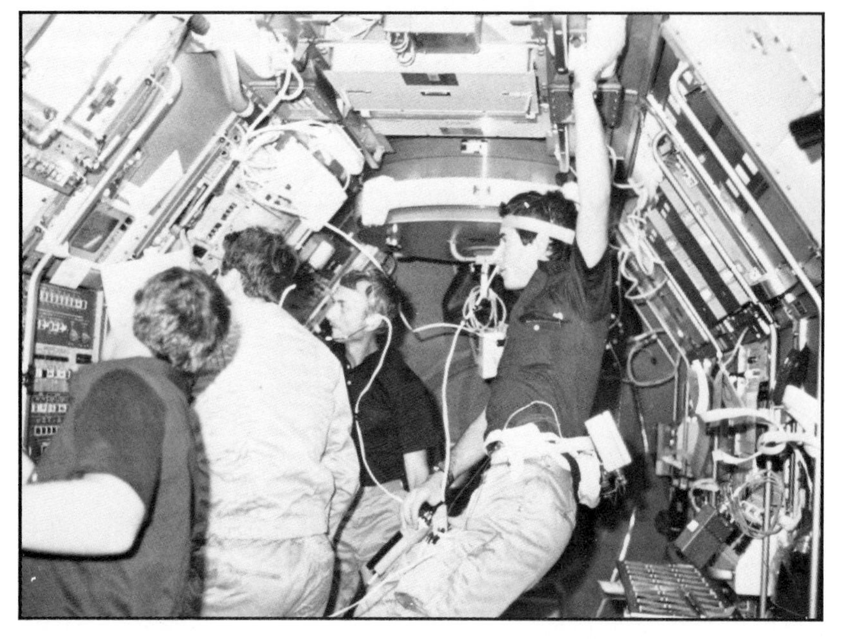

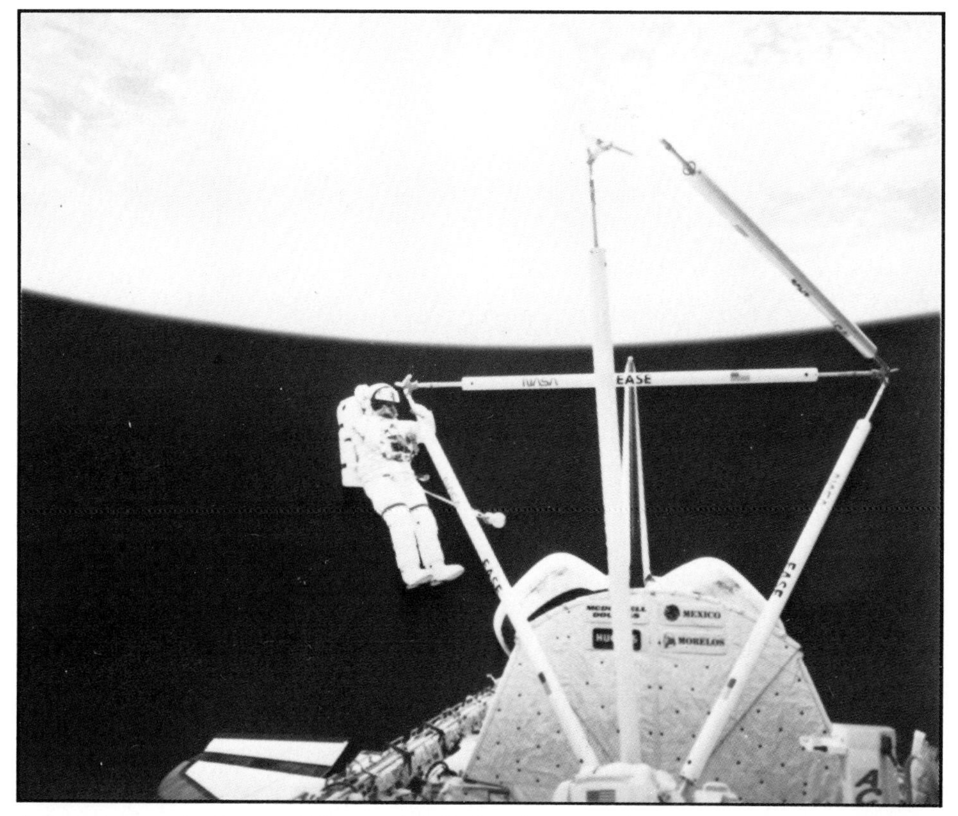

Astronauts Jerry L. Ross and Sherwood C. Spring shown working with the EASE/ACCESS experimental structure during STS 61-B. Also present on this Shuttle mission was McDonnell Douglas employee Charles D. Walker, who operated the company's experimental pharmaceutical production apparatus.

View of the EASE/ACCESS experimental construction beam, deployed during STS 61-B in November 1985. This experiment provided data to help design and develop the main truss structure of NASA's Space Station.

An experimental panel deployed from the Space Shuttle during a mission in 1985. Due to the lack of gravitational forces, extremely thin structures can retain their rigidity in space.

(Top) A composite photograph of astronaut Harrison H. Schmitt and the lunar landscape, taken during NASA's last mission to the Moon — Apollo 17 — in December 1972. Lunar samples returned to Earth by the Apollo astronauts show evidence of abundant sources of usable raw materials on the Moon.

(Bottom) The trajectory stabilization segment of a lunar mass driver is shown in this artist's concept. Small bundles of raw lunar material can be directed from this device toward a mass catcher in orbit around the Moon.

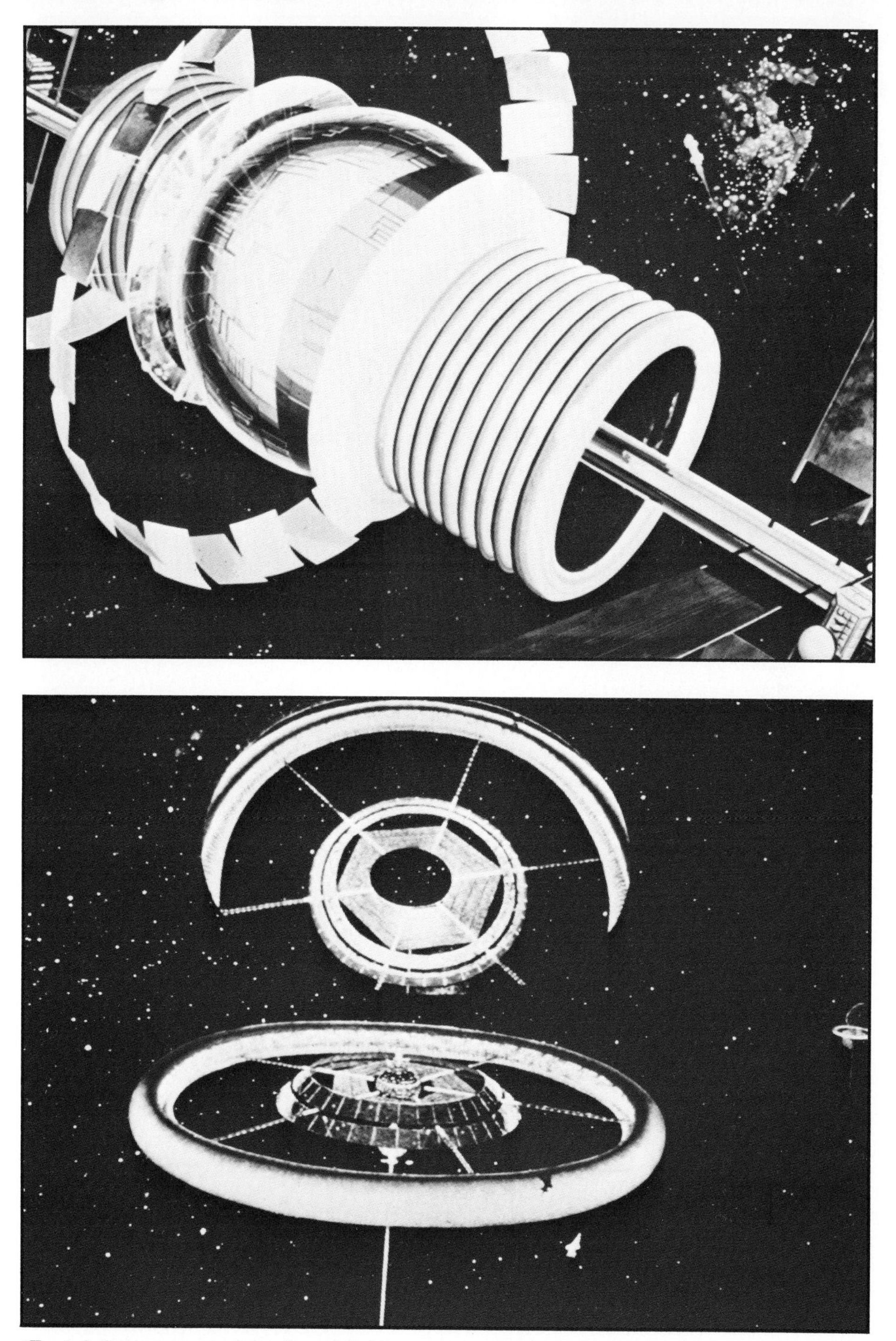

(Top) Artist's concept of the Bernal Sphere, one proposed design for a space colony housing ten thousand people. As conceived by Dr. Gerard K. O'Neill, this habitat is roughly one mile in diameter and has an Earth-like interior resembling a small college campus.

(Bottom) A toroidal space settlement that resembles the "rotating wheel" space station concept popularized by Wernher von Braun in the 1950s. In this artist's concept, natural sunlight is reflected into the 100,000-person habitat by a large orbiting mirror.

(Top) An artist's concept of the interior of a toroidal space colony. Space habitats such as these would be rotated slowly to provide their residents with artificial gravity.

(Bottom) The interior of a cylindrical space colony capable of housing up to ten million people is shown in this artist's drawing. A settlement of this size could probably be built by the middle of the next century with technologies currently being developed.

III

Beyond the Gateway

One of NASA's greatest shortcomings is the space agency's lack of experts who combine a practical knowledge of current space systems with a vision of humanity's long term future in space. Most government and industry personnel who are involved on a daily basis in the design and operation of today's major space systems, most notably the Space Shuttle and the Space Station, are predominantly concerned with demonstrating and achieving the near-term benefits required to sustain political and economic support for these programs. Long range goals, such as space industrialization and interplanetary travel, have little impact on the activities of these individuals. In the bureaucratic offices where space policy is developed, there is a widespread perception that the benefits of long range programs are too remote to justify their incorporation into the mainstream planning cycle.

Meanwhile, very few space visionaries seem to have any impact on the day-to-day activities that guide the course of America's Space Program. Ambitious space activities, many of which require more than ten years of development and ten billion dollars for completion, have remained the exclusive realm of storytellers and "wild-eyed enthusiasts" since the end of the Apollo era. Advocates of large scale space development are generally kept at arm's distance by NASA's upper management, and have received little official support. NASA has spent one hundred billion dollars over the past fifteen years, but less than one-thousandth of this money has been devoted to post-Space Station goals. The agency's only significant effort to investigate a long range goal was a $20 million study of solar power satellites co-sponsored with the U.S. Department of Energy during the 1977-1980 time frame. Even this effort was fatally flawed by lack of vision; the study was conducted with the limiting groundrule that all satellite construction materials would be launched from Earth. By ignoring the potentially attractive option of obtaining materials from the Moon or other nonterrestrial sources, the participants in this study virtually assured that the project would be found environmentally and economically unacceptable.

In 1983, President Reagan's science advisor George Keyworth criticized NASA for lacking long term vision, testifying before Congress that NASA's Space Station concept would be an "unfortunate step backward." Keyworth later softened his opposition to the Space Station, but insisted that NASA outline a grander scheme for space development and demonstrate how the Space Station fit into this overall plan. At the same time, however, NASA was being pressured by others within the White House to settle on a concept for an austere Space Station that could be developed with little or no growth in the agency's budget. In this instance, as in the case of the Space Shuttle one decade earlier, NASA's ability to develop a long term space development strategy was hampered by a familiar foe: budgetary pressure. As early as 1963, NASA's budget requests were trimmed by Congress or the Executive Branch of the government.

Affordability, rather than technical feasibility or long term utility, is the principal criterion used by the U.S. Government to evaluate new space programs.

The Keyworth incident also illustrates a second factor that has made it difficult for NASA to establish long range objectives: disagreement within the government as to what these objectives should be. Keyworth's main objection to the station was that NASA designed it to be permanently manned. Many individuals within NASA and the rest of the aerospace community share Keyworth's view, believing that unmanned space exploration should take precedence over the more expensive manned programs. But the desire of NASA's leadership to develop a fully *manned* station was unequivocal from the outset, and the agency's Space Station contractors loyally followed suit. Until a consensus is reached on man's role in space and how the Space Station should support this role, broad support for a long range space development strategy will be difficult to achieve.

Responding to the need for the United States to develop a cohesive long range space policy, the National Commission on Space solicited ideas from hundreds of American citizens in the preparation of its 1986 report, *Pioneering the Space Frontier*. The commission proposed a dramatic expansion of the nation's unmanned space science program, but made quite clear its finding that our fundamental objective should be to build a home for humanity in space. Its report states that "the future will see growing numbers of people working at Earth orbital, lunar, and eventually, Martian bases, initiating the settlement of vast reaches of the inner solar system." The principal recommendation of the commission was that this expansion should be well underway within the first two decades of the twenty-first century.

The long range goals espoused by the National Commission on Space are not often discussed in public by NASA officials, who have a sufficiently difficult time keeping modest short term projects funded. But as the commission pointed out, these longer range programs can be achieved with today's technologies or technologies that can be developed within the next few decades. Their costs are within the range of what can be afforded with about the same level of commitment the Space Program enjoyed during the 1960s. The challenge facing us today is to justify this renewed commitment by developing an integrated plan that takes us from today's reality to tomorrow's vision. Such a plan must achieve these grand objectives in an expedient manner, resulting in tangible benefits that can be enjoyed by a large share of our planet's population within the shortest possible time.

7

The Role of Man in Space

One of the complaints most frequently levied against NASA is that the United States Space Program lacks a cohesive, long range plan. Many observers believe that a master strategy for space development is needed if our space activities are to have meaningful and lasting benefits. A broad plan would provide unity of purpose, helping to ensure that our various space projects complement each other. It would define an ambitious and fruitful objective, helping to sustain public enthusiasm and support. Perhaps most important, it would help everyone understand why we have a Space Program, and give us a clear measure of success in evaluating our long term progress in space. Development of a unified national space policy has been difficult to achieve because the thousands of people who collectively set our space goals have never been able to agree on a single course of action. Even near term goals designed to have broad appeal, such as NASA's Space Station, encounter opposition from large numbers of people who may otherwise be supportive of the Space Program. The basis for much of this diversity is disagreement over one fundamental issue: the role of man in space.

Man in Space Before the Space Station

In the years preceding approval of the Space Station Program, the role of humans in space was easily defined. During NASA's formative years, the era of the "original seven" astronauts selected for Project Mercury, the space agency's sole manned space flight goal was to demonstrate that human beings could survive for short periods of time and perform basic functions in outer space. Shortly after the first Mercury flight, President Kennedy extended this goal by selecting a round-trip to the Moon as the preeminent goal of the entire 1960s manned space program.

Kennedy's ambitious goal ensured that America's Space Program would be dominated by manned space projects for nearly a decade. Space agency officials endorsed this approach wholeheartedly, largely because manned space missions evoked tremendous public interest, helping to foster continued political support for all of the agency's programs. NASA's manned space flights received far more publicity than the practical accomplishments that were being made by unmanned spacecraft in the areas of telecommunications, weather forecasting, and space science. The benefits of early unmanned space programs were substantial, but it was the human drama of manned voyages into the unknown that attracted spectators to launches, drew national television audiences, and resulted in multibillion dollar budgets. If not for the human element of the Mercury, Gemini, and Apollo missions, the public may not have given so much support to the Space Program and NASA's rapidly growing expenditures.

After President Kennedy's dream was realized and the excitement of NASA's Moon shots began to wane, the role of man in space needed to be redefined. But developing a new concept of what people should do in space turned out to be an unexpectedly difficult task, so the issue was largely sidestepped by America's political leadership. Of the six major national space objectives outlined in President Nixon's 1970 space policy, only one dealt specifically with this subject, and its mandate was vague: "to extend man's capability to live and work in space." Not one major program was established in response to this policy objective. In fact, NASA's Skylab project and a more advanced space station program, which were already in progress and dedicated to achieving this very objective, were both curtailed to save money within a year of Nixon's policy statement.

Nonetheless, the Skylab station was extremely successful in extending human capabilities in space. After the station was severely damaged during its unmanned launch in May 1973, the Skylab I crew visited the station and manually deployed a sun shield and solar array, salvaging the entire sequence of three Skylab missions. The Skylab I astronauts and two subsequent crews performed a total of five thousand hours of scientific observations and experiments, setting the stage for some of the most significant Space Shuttle accomplishments of the 1980s. But the Skylab program ended with the station's unfortunate destruction in 1979, largely as a result of NASA's failure to map out and support a long range strategy for extending Skylab's mission.

By 1976, NASA's entire manned space flight program was devoted to development of the Space Shuttle, conceived in 1971 in response to a different Nixon policy objective: "to substantially reduce the cost of space operations." By interpreting this as a directive to reduce the cost of *manned* space operations as well as unmanned activities, NASA was able to justify the economic drawbacks of making the Shuttle a manned system. When the Shuttle became operational a decade later, its initial successes appeared to vindicate NASA's preference for manned systems. Despite high operating costs, the Shuttle recaptured the level of public enthusiasm that had existed during the Apollo and Skylab years, giving an enormous boost to the Space Program. The Shuttle also helped demonstrate, as had Skylab, that manned spaceflight has certain unique advantages. Man's utility in space was showcased in especially dramatic fashion during the April 1984 mission to repair the Solar Max satellite, an unmanned solar observatory that had been malfunctioning for several years. By snaring the Solar Max satellite in an unplanned manner with the Shuttle's Remote Manipulator System, the astronauts demonstrated the ability of humans to cope with unforeseen circumstances in ways that machines cannot.

But the Challenger accident brought America's manned space program to a sudden halt and raised new questions about the costs and benefits of manned space flight. There is now an emerging realization that the nation's major space programs were selected and implemented throughout the first thirty years of NASA's existence without a guiding consensus on the long term role of humans in space. The desire to progressively expand man's capability to perform routine tasks in space certainly influenced all manned projects from Mercury through the Space Shuttle, but America's space planners never reached a broad agreement on a set of long range objectives for manned space flight. To accomplish this, Space Program officials must address the fundamental issue of whether humans should ever live in space on a *permanent, ongoing basis*. Since all manned projects through the Shuttle era were designed for short duration flights, the need to decide whether space is an extension of mankind's natural home rarely confronted NASA's leadership. But with the beginning of our transition into the Space Station era, we will soon need to identify the long term benefits of manned space flight and develop a comprehensive strategy for achieving these benefits.

The Space Station and the Role of Man in Space

By committing the United States to the development of the Space Station, President Reagan endorsed NASA's goal of moving on from the Shuttle to establish a "permanent manned presence" in space. This was a significant victory for NASA's manned space flight advocates, and a defeat for the sizeable faction within the aerospace community that opposes manned space programs. Opponents of manned space flight believe that the benefits of space development can best be realized with unmanned spacecraft such as satellites, platforms, and planetary probes. Manned space projects, they argue, are much more expensive than unmanned activities, and do not provide sufficient benefits to justify their additional cost.

Advocates of unmanned space exploration are generally frustrated at what they perceive as a NASA bias toward manned space projects. In their view, NASA's manned space programs have diverted tens of billions of dollars from unmanned projects that could have provided greater benefits, at lower human and economic costs. One of the leaders of the unmanned space advocacy is Dr. Bruce Murray, who headed the NASA/California Institute of Technology Jet Propulsion Laboratory during the heyday of unmanned space exploration in 1976-1982. Dr. Murray summed up the feelings of his constituents in early 1986 by stating that "manned space flight is justified by political but not utilitarian reasons." The short history of our Space Program lends some credence to Dr. Murray's argument. Today's manned space programs are comparatively expensive, owing to the complexity and high cost of systems designed to keep people alive in the hostile environment of space. These economic costs are reflected by NASA's annual budget, which consistently provides manned space programs with four to five times more funding than unmanned projects. And despite their costly safety precautions, such projects are inherently risky to their human cargo, as demonstrated by the devastating loss of the Shuttle Challenger and her crew.

Not only are the costs and risks of manned space flight great, but most quantifiable benefits of the Space Program have resulted from our unmanned space projects. Satellite communications, weather forecasting, and remote sensing of Earth's resources have all been routinely facilitated by unmanned spacecraft since the 1960s. Many scientific and other intangible benefits have also been obtained through the use of unmanned systems. Deep space probes such as Pioneer and Voyager have returned incredible amounts of information and dramatic photographs to Earth, helping us to experience regions of our solar system far beyond the reach of today's manned vehicles.

An argument frequently offered in favor of manned spaceflight is that the human experience itself is an important element of space exploration. No space-borne machine, it is argued, could possibly bring back to Earth the same insights as humans experiencing space travel first-hand. This is certainly an intangible benefit, but is it fair to assert that manned space projects do not provide any utilitarian value? The eight aerospace contractors that performed NASA's Space Station mission analysis studies in 1982 and 1983 apparently believed that there are a number of sound reasons for having people live and work in space. They concluded unanimously that the core facilities of NASA's proposed Space Station should be permanently manned, with co-orbiting unmanned platforms representing a relatively small share of the overall Space Station system.

Man's presence in orbit was determined to be particularly vital for materials processing in space research, owing to the fact that MPS projects are begun with exhaustive trial-and-error experimentation. McDonnell Douglas has determined that it does not need the station to support its commercial electrophoresis venture, but the company's experience conducting MPS research on the Space Shuttle confirms the value of a "man-in-the-loop" during the early phases of MPS experimentation. To monitor experiments and to make adjustments when necessary, there is no substitute for a human presence.

The mission analysis contractors concluded that humans will be even more indispensable to the achievement of other Space Station objectives. To conduct experiments aimed at understanding the effects of the zero-gravity environment of space on plant and animal life, at least two or three Space Station crewpersons will have to be assigned full time to life science experimentation. NASA and the agency's Space Station contractors also believe that a human crew will prove valuable for adjustment and repair of unmanned space science experiments located on or near the Space Station. Based on these requirements, NASA's current plans call for the Space Station to be inhabited initially by a crew of six persons, with expansion of the crew size as Space Station operations grow in scale and complexity. In total, NASA expects that as many as twenty men and women will be living and working at the Space Station within the next fifteen to twenty years.

These plans have been challenged by the unmanned space advocacy, which views the manned Space Station as another money-absorbing project that will reduce the amount of funding available for unmanned space exploration. Many space scientists are particularly critical because they personally have devoted their careers to unmanned programs that were scaled down or cancelled by NASA to make money available for completion of the Space Shuttle. They fear that the Space Station will have a similar effect, and would rather see the unmanned elements of the station developed first. This option, they argue, would accomplish many space science goals earlier and at a lower cost than would be achievable by going directly to a manned capability. Some opponents of the manned Space Station contend that an unmanned station would also do more to advance the technologies of automation and robotics, resulting in a greater number of "spin-off" benefits on Earth.

An alternative Space Station proposal that emphasizes unmanned elements is the "Pleiades," or "Space Flotilla" concept developed in 1983 by scientists and students at Stanford University. The Space Flotilla would consist of a constellation of unmanned instruments and experiments that would be deployed and periodically revisited by the Space Shuttle. In response to pressure from Pleiades advocates, NASA directed its Space Station contractors in 1984 to consider a "man-tended" alternative to the permanently manned Space Station. However, space agency officials clearly preferred the option of having the Space Station permanently manned from the outset. NASA selected a permanently manned "reference configuration" as the design basis for its definitive *Space Station Concept Definition and Preliminary Design* studies, and the agency specified that man-tended designs were to be given only one-tenth as much emphasis as manned versions of the Space Station.

In defending their position, NASA officials argue that space operations, including those currently accomplished with unmanned systems, will increase in complexity and will eventually require human interaction. By supporting a human crew from the outset, the Space Station can meet all such requirements as they emerge. Availability of the Space Station, NASA contends, will also encourage future users of the space environment to design their equipment to take full advantage of man's unique capabilities. As an example, the Space Station might be used for assembly and servicing of large geostationary platforms, whose development would be less likely without the availability of space-based construction and maintenance crews.

Beneath the surface of these arguments lie even more fundamental reasons for having people in space, some of which are not often discussed publicly by NASA's leadership. The space agency usually focuses on near term utility when defending its approach to Space Station development, but the long range implications of human expansion into space are an important underlying consideration. Many Space Station advocates would support the development of a manned facility even if every one of its Earth-serving functions could be performed more efficiently in an automated manner. All of NASA's manned programs have been supported by such people, who believe it is mankind's ultimate destiny to populate regions beyond the reaches of our home planet. To these individuals, a manned Space Station is the most logical next step in the inevitable expansion of the hospitable realm of our Universe.

Humanity in Space: Beyond the Space Station

People have fantasized about living on the Moon, on Mars, or in imaginary places of an un-Earthly nature for ages. Today, more people than ever are captivated by the concept of space travel, a prospect that is suddenly not as far-fetched as it was only one or two generations ago. Many such individuals are motivated by the same factors as their counterparts in previous eras: the desire to explore the unknown, to seek new experiences, or to fulfill other intangible yearnings. The urge to explore and settle new frontiers appears to be a deep-rooted human characteristic, and the Space Program has always derived support from people who believe that this alone is sufficient justification for providing increasing numbers of people with access to space.

During the early 1970s, however, a modern goal-oriented agenda for the large scale human settlement of space began to emerge. This concept, popularly known as *space colonization,* gained widespread popularity shortly after the 1972 publication of *The Limits to Growth,* the first in a series of books forecasting the demise of human civilization. The basic premise of *The Limits to Growth* was that life on Earth will become progressively more miserable, due primarily to three factors: overpopulation, diminishing resources, and pollution of the environment. This book and its sequels helped catalyze the embryonic space colonization movement, led by Princeton physicist Gerard K. O'Neill and numerous colleagues around the country.

O'Neill's response to *The Limits to Growth* is his 1976 book, *The High Frontier,* which presents his vision of Earth-like space colonies capable of housing tens of millions of people by the early twenty-first century. In O'Neill's visionary scenario, ninety percent of the building materials required to establish space settlements would be obtained from nonterrestrial sources, primarily the Moon and the asteroids. An endless supply of solar energy would be tapped to provide food and power for all of humanity. All polluting industries would be moved into space, preserving Earth's delicate biosphere forever. According to O'Neill's calculations, the resources of the solar system could be used in this manner to support a population of eighty *trillion* people.

O'Neill's vision was particularly appealing to people who rejected the bleak and stifling conclusions of The Limits to Growth. It also provided a very practical argument for manned spaceflight, as long as one maintained a sufficiently long range perspective. NASA endowed O'Neill with a modest amount of funding for three summer studies, which convened dozens of government, industry, and academic experts to evaluate the feasibility of space colonization in 1975, 1976, and 1977. The participants in these studies unanimously concluded that all of the technologies required for large scale space habitation were available or could be developed within the span of two or three decades. The last of these studies, conducted at NASA's Ames Research Center during the summer of 1977, outlined a plan for establishing large space settlements by the turn of the century. The first key milestone in this plan: use of the Space Shuttle to assemble a manned Space Station in low Earth orbit by the late 1980s. While NASA's current Space Station plans do not officially cite space colonization as one of the missions for this facility, many NASA people will admit during candid moments that they personally believe that the Space Station's ultimate objective is to make visions such as O'Neill's a reality.

The value of expanding the human presence in space is clearly dependent upon the basic premises one makes regarding the relationship of mankind to his home planet. In its 1986 report, the National Commission on Space states that "long term exponential growth into eventual permanent settlements should be the overarching goal" of America's Space Program. If the role of the Space Program is to be limited to enhancing national security and providing people on Earth with minor conveniences and amenities, then such a goal would be difficult to justify, because unmanned spacecraft could probably perform a vast majority of the work to be done in space. Communications, remote sensing, certain types of scientific exploration, and even small scale materials processing can all be accomplished with automated space systems. But with the technological advances of the last two decades, manned space programs may offer the additional opportunity to make dramatic improvements in the quality of life for a significant number of Earth's inhabitants. To feed, clothe, educate, and entertain a growing population presents a challenge that may be met by utilizing the resources of the solar system, which will require a growing, self-supporting human presence in space.

The space colonization movement has helped demonstrate that there may in fact be utilitarian reasons for giving large numbers of people the opportunity to live and work in space. An argument frequently presented against space colonization is that the improvements in quality of life sought by space enthusiasts might be achieved at lower cost through conservation, population control, and use of terrestrial-based technologies. Many space enthusiasts accept this premise, but point out that the latter approach requires that we intentionally limit the sphere of human activity and accept growth constraints that we may have the technology to defy. At issue is whether we should seek the most expedient solutions to our problems, or whether we should try to meet our material needs in ways that also satisfy intangible needs, such as the desire to experience new places and new lifestyles. Space development may not represent the most direct way of meeting material needs, but arithmetic cost-payback ratios will never capture the full essence of what attracts humans to space.

In developing our long term strategy for expanding the human presence in space, we must balance our desire for rapid material and economic payback with the realization that not all human endeavors need to be justified in narrow utilitarian terms. If the various benefits that we hope to achieve are clearly defined, then the unified long range plan that we have lacked for so long may finally be established. Supported with the proper balance of motivation and the conscious pursuit of long range goals, the Space Program has the potential to enrich our lives in ways that could not be achieved through any other means. If the migration of people into space is indeed a natural evolutionary step, then the benefits of space habitation should be long lasting and the concept of mankind's permanent home will have to be expanded, to include the solar system and perhaps much more.

8

Firing Up the First Space Factories

When the first European settlers arrived on this continent five centuries ago, they immediately set to work to transform the natural resources of their new homeland into the usable products they needed to survive. America's new inhabitants had to be self-sufficient from the moment they set foot on New World soil, for they certainly could not have imported their food, clothing, or building materials from established sources overseas. It was gold, spices, and other exotic materials that initially attracted explorers to the American continent, but it was the abundance of simple necessities such as water, good soil, and timber that enticed visitors to remain in the New World permanently. Similarly, human excursions into space will evolve into a permanent migration from Earth only after we develop the ability to convert the resources of the space environment into the basic elements of survival. NASA's Space Station will support exotic functions that will justify short duration visits into space, but the Space Station must logically be followed by development of an infrastructure for using space resources: a base of industrial production in space.

The Need for Space Resources

Space station residents could live on "care packages" from Earth and recycled waste products for many years, but continued reliance on Earth would be extremely expensive. In fact, the cost of resupplying the Space Station crew with such necessities as air, food, and water could represent up to 90% of the total cost of operating the Space Station. Even with a relatively modest initial crew of six persons, the transportation costs associated with resupplying the Space Station will exceed $1 million per day. The cost of delivering material to the station may decline as more cost-effective Earth-launch systems are developed, but this benefit will be offset as the size of the Space Station crew is expanded. And as the human presence is expanded beyond low Earth orbit, the cost of maintaining supply lines between Earth and our space outposts will become even more prohibitive. The cost of delivering cargo from Earth to resupply facilities on the Moon, Mars, or in high orbits will be several times more expensive than replenishing the LEO Space Station. Even with very advanced space transportation systems, it will probably cost more than ten thousand dollars per pound to launch materials from Earth to these outer destinations. At the same time, the cost of manufacturing items in space will decline as human activity moves closer to potential extraterrestrial sources of raw materials such as the Moon and the asteroids.

For the use of nonterrestrial resources to become economical, however, the need for these materials must be sufficient to justify the billions of dollars it will cost to establish space mining and manufacturing facilities. The Moon is the most likely site for an economical space manufacturing enterprise, since we succeeded in staging short duration human journeys to the lunar surface with technologies that are now two decades old. From the eight hundred pounds of lunar soil and rock returned to Earth by the Apollo astronauts, we know that the Moon is rich in a number of elements that will be in high demand as the population in space expands. These include oxygen, which is the principal ingredient in rocket propellants as well as an obvious necessity for life support. The Moon also contains an abundance of aluminum, which can be used as the primary building material for large space structures, and silicon, a major component of solar cells and glass. Since the gravitational pull of the Moon is only one sixth as great as that of Earth, raw materials can be delivered from the Moon to Earth's orbit with less than one twentieth of the energy that would be required to deliver the same quantity of material from Earth.

Many important elements that cannot be obtained from the Moon, such as nitrogen and hydrogen, can ultimately be derived from other near-Earth sources, such as Mars, Martian moons, and asteroids. Once materials from these sources are delivered into space, they can be moved about with relatively little consumption of energy because they are essentially weightless. The energy that is required for transportation of raw materials may eventually be obtained by the sun, through use of solar electric propulsion vehicles. The uninterrupted availability of intense solar energy in space can also be used to power the factories that process these materials. Another potential source of power is nuclear energy, which can be utilized in space without the environmental concerns that restrict its application on Earth. Space factories will also be able to operate unimpeded by the multitude of other environmental concerns that must be considered by industrialists on Earth.

Since the early 1970s, NASA, aerospace firms, and universities have conducted numerous studies and experiments that have confirmed these basic principles. Most experts on space development now agree that large scale space manufacturing is technically feasible and could be accomplished by the beginning of the next century. But identifying a use for space resources that offers a near term return on investment remains a challenge. Even a relatively simple production plant on the surface of the Moon would cost several billion dollars to design, launch, and assemble. Such an enterprise could be justified on economic grounds only if it promised to provide large quantities of material at a substantially lower cost than if these materials were acquired from Earth. The more materials that are required to support space activities and the farther from Earth they are needed, the greater the chance that it will be economically attractive to obtain these materials from nonterrestrial sources. According to recent NASA studies, about one million pounds of such products would be needed per year in low Earth orbit for a production facility on the Moon to be operated economically.

For this reason, the earliest economical application of nonterrestrial resources will probably be the use of oxygen derived from the Moon for orbital transfer vehicle propellants. The composition of lunar soil is approximately forty percent oxygen by weight, leading some space experts to characterize the Moon as a "tank farm in space." According to current projections, the OTV's based at NASA's Space Station will consume about one million pounds of cryogenic propellants — liquid hydrogen and oxygen — each year, beginning around the year 2000. Delivered from Earth, these propellants would cost only about one dollar per pound to manufacture, but would cost at least one *thousand* dollars per pound to launch to the Space Station. Alternatively, if OTV's could operate routinely between the Space Station and the Moon, propellants might be brought from the Moon to the Space Station for as little as $50 to $100 per pound, saving about one billion dollars per year in propellant transportation costs. Benefits of this magnitude could justify the expense of developing a lunar base capable of extracting oxygen from lunar ore.

Through its studies of space resource utilization, NASA is seeking to develop an accurate estimate of how much it would cost to develop and operate an early lunar base that could manufacture liquid oxygen. Such a facility would have to support a permanent lunar base crew, a power system capable of providing several thousand kilowatts, and equipment for mining and processing lunar sand and rocks. Based on current indications, it appears that a small scale initial lunar base might be established for about the same cost as the Space Station, or about ten billion dollars. This is a substantial investment, but it would provide a permanent source of low cost materials for a fraction of the outlay required to stage the Apollo missions to the Moon in the late 1960s and early 1970s.

After achieving the capability to produce oxygen on a routine basis, an initial lunar base could be expanded to manufacture other materials. Extracting from lunar soil the other constituent of cryogenic rocket propellant — hydrogen — might be technically feasible, because tons of hydrogen have been deposited on the surface of the Moon by solar particle bombardment over the past few billion years. But at concentrations of only fifteen parts per million, lunar hydrogen may still be too rare for its extraction and production on the surface of the Moon to be economical. With greater likelihood, the expanding lunar base would be used for production of aluminum and silicon, both of which are abundant in lunar ores.

Conceivably, either aluminum or silicon could be used as a rocket fuel, replacing hydrogen fuel in advanced engines. It is more likely, however, that such elements will be mined from the Moon for application in the construction of large space structures.

Aluminum could be employed in the assembly of large space platforms, which may be needed by the early twenty-first century for advanced communications systems or "Star Wars" missile defense systems. Aluminum and silicon could also become the primary elements used in the construction of solar power satellites, which NASA and the Department of Energy studied in the late 1970s as a potential means of beaming solar-generated electricity to Earth.

On the more distant horizon, use of materials mined from the Moon will eventually feed the space factories that enable mankind to populate space on a truly large scale. Permanent space settlements such as those envisioned by Gerard O'Neill and his contemporaries could not be established without the availability of millions of tons of raw materials mined from nonterrestrial sources such as the Moon and the asteroids. The extensive NASA/ Department of Energy studies of solar power satellites concluded that such a satellite power system could not be manufactured from Earth materials due to economical and environmental factors, and space colonization would require much more raw material than a satellite power system. The economics of near term space manufacturing remain uncertain, but there is no question that we will ultimately need an industrial base in space if our population is to move beyond the bounds of Earth.

A Practical Approach to Space Industrialization

The development of space manufacturing technologies will be made possible by the growth of materials processing in space and satellite communications, industries that will be given a big boost by the Space Station. Microgravity processing will require advancements in techniques for acquiring solar power and carrying out automated production tasks in space. These capabilities will first be demonstrated on small scale space processing projects at the Space Station in the mid-1990s, and will grow in sophistication as the market for space-processed materials expands. Once these technologies are demonstrated, they can subsequently be adapted to larger scale processing activities utilizing nonterrestrial resources. The satellite communications industry will help make space manufacturing possible by stimulating the development of another enabling technology: advanced orbit-to-orbit transportation systems. By the turn of the century, orbital transfer vehicles will routinely ferry communications satellites from the Space Station to geosynchronous orbit. These same vehicles will be capable of supporting large scale space manufacturing by hauling factory supplies and raw materials from the Earth and Moon to orbiting factories.

The Space Station will also provide hardware and technologies that will reduce the investment required to establish human settlements on the Moon. It will be possible to modify Space Station habitat modules and deliver them to the Moon to house a lunar base crew. Use of modified Space Station life support accommodations would probably cost several hundred million dollars less than developing a completely new design for lunar base habitats. Lunar base operations costs might be reduced if Space Station automation and robotics techniques could be applied to the tasks of mining and processing lunar ore, enabling these jobs to be accomplished with minimal crew requirements. Recognizing the importance of the Space Station and lunar production to our long term space objectives, NASA has sponsored studies to develop detailed plans for expansion of the human presence from low Earth orbit to the Moon. These studies have indicated that once a Space Station is established, a large scale lunar production plant and orbiting space factories could be developed within the span of approximately one decade.

These studies have also generated the important conclusion that basic elements of the Space Shuttle, as well as the Space Station, could be employed to reduce the costs and time required to establish an industrial base in space. The Space Shuttle's huge external fuel tank might some day serve as a particularly important building block for large scale space activities. The performance of the Space Shuttle would actually be enhanced if the external tank were brought into space with the Orbiter, since the current practice of expending the external tank over the Indian Ocean requires the Shuttle to execute an inefficient orbital maneuver.

Once in orbit, the external tanks could be used for a variety of purposes, saving the enormous cost that would be involved in launching the 80,000 pound weight of each of these tanks to orbit as standard payload on the Space Shuttle or expendable launch vehicles. According to NASA studies, a single two hundred foot long, twenty-seven foot diameter external tank could be outfitted in space to serve as a habitat for as many as twenty-one persons, each of whom could have a private "apartment" with nearly two hundred square feet of living space. By clustering these tanks together, several hundred people could be housed at one facility, enabling truly large scale space enterprises to be undertaken. One NASA concept even entails connecting several tanks with cables and rotating the configuration slowly to provide artificial gravity for its inhabitants.

By using technologies and hardware derived from the Space Shuttle and Space Station, human communities of this scale could be established in space by the early twenty-first century without major increases in space expenditures. Facilities employing similar designs could be set up in low Earth orbit, on the lunar surface, and in stable Earth-Moon orbits, providing a space manufacturing network capable of using nonterrestrial materials on a very large scale. The first detailed plan for establishing such a network was developed four years before the Shuttle became operational and is presented in NASA's pioneering 1979 report, *Space Resources and Space Settlements*. This volume documents the work performed by eighty leading experts on space industrialization, who were convened by NASA during the summer of 1977 to put together a plan for rapid achievement of large scale space manufacturing. The participants in this study, which was directed by Gerard O'Neill, concluded that a large industrial base could be established in space within this century. Their report provides a step-by-step plan for developing and operating such a system and defines the basic scientific and engineering requirements to conduct this type of program.

The focus of the NASA space manufacturing scenario is a large orbiting factory referred to as the Space Manufacturing Facility (SMF), which is located in a stable orbit equidistant from the Earth and the Moon. The first step toward establishment of the SMF would be the assembly of a work station in low Earth orbit, consisting of six refurbished Shuttle fuel tanks and accommodating up to eighty-four crew members. A total of sixteen Shuttle flights would be required to deliver the elements of the LEO space station into orbit. When NASA originally developed this concept, the agency expected that the Space Shuttle would be flown once every week, so the time required to assemble the LEO station was projected to be about four months. However, our experience in operating the Space Transportation System indicates that this task would actually take significantly longer, since the sixteen Shuttle flights needed to deploy the LEO station would probably have to be spread over a period of at least two years.

After completion of the LEO space station, the next step in this scenario is for the Space Shuttle to deliver the elements of a lunar mining base to low Earth orbit. The LEO crew would assemble the lunar base habitat, which could also be derived from a

Shuttle external tank. Other elements of an initial lunar base would probably include a bulldozer and a truck to gather lunar ore and a "mass driver" to launch this ore from the lunar surface. The LEO crew would load the lunar base elements onto a large cargo-carrying orbital transfer vehicle, which would deliver this payload to the surface of the Moon. Once this were accomplished, a much smaller manned OTV would carry an initial lunar base crew of up to twelve persons from LEO to their new work station on the Moon. This crew would be responsible for final assembly of the lunar base elements and would manage the largely automated lunar mining operation.

The mass driver, which is essential for the cost-effective delivery of lunar materials into space, represents the greatest technological challenge of this space manufacturing scenario. As envisioned by NASA, the entire mass driver facility would be over one hundred miles in length and could ultimately launch over 600,000 tons of lunar ore into space each year. To provide the energy needed to catapult this ore beyond the gravitational pull of the Moon, the mass driver would employ a three hundred meter long array of electromagnetic coils and would accelerate magnetically charged buckets of lunar ore to velocities of over 20,000 miles per hour in a matter of seconds. The buckets would then pass through a mile-long series of guide rails, after which they would be deflected and returned to the accelerator section. The "slugs" of lunar ore would continue on toward trajectory stabilization stations several miles downrange, which would aim the ore toward a "mass catcher" device in orbit about the Moon. Efforts to develop the technology for a lunar mass driver are currently being given a boost by President Reagan's Strategic Defense Initiative program, which involves multimillion dollar research on orbiting electromagnetic accelerators.

During start-up of lunar base operations, the crew back at

LEO would turn their attention to assembly of the Space Manufacturing Facility, whose elements could also be delivered from Earth via the Space Shuttle. The design of the SMF would probably be similar to that of the LEO station, employing refurbished Shuttle fuel tanks as basic structural elements for habitats and work areas. As envisioned by the participants in NASA's 1977 summer study, the SMF would have three main sections. The primary section would be a chemical processing plant for converting lunar ore into desired elements, primarily aluminum, silicon, iron, titanium, and oxygen. Aluminum could be obtained from lunar anorthite through a dual process involving carbochlorination and electrolysis. Iron and titanium could be removed from lunar ilmenite through carbochlorination and subsequent calcium and hydrogen reduction. Both of these metal extraction processes would generate large amounts of usable carbon and oxygen as by-products.

The second section of the Space Manufacturing Facility would be a fabrication plant to transform these elements into the end products of the space manufacturing enterprise. The energy to run the plant would be provided by the third major element of the SMF, a power generation system consisting of photovoltaic arrays and radiators. The participants in NASA's 1977 study were directed to design a Space Manufacturing Facility capable of fabricating solar power satellites, which would transmit energy to Earth in the form of microwaves. Each of these satellites would be several miles in length and have a generating capacity of ten million kilowatts, enough power to meet the electrical needs of a large city. To meet the study goal, which was the production of 2.4 such satellites per year, it was projected that the SMF would have to support a crew of three thousand persons.

To support this many people and such a high level of production, it was estimated that the SMF would have to grow from an initial mass of approximately two thousand metric tons to a full scale mass of about thirteen thousand tons, more than half of which would be delivered from Earth. Since several hundred launches of the Space Shuttle would be needed to deliver this quantity of material to LEO, a heavy lift launch vehicle, with up to three times the payload capability of the Shuttle, would probably be required to efficiently support an enterprise of this scale. Advanced forms of interorbital transportation would also be needed. NASA's space manufacturing scenario involves the use of large interorbital cargo vehicles whose design is based on that of the lunar mass driver. The "reaction mass" fuel for these "mass driver reaction engines" (MDRE's) would be solid particles such as pellets made from ground-up Shuttle fuel tanks or the waste products of space manufacturing. During a single four month journey, one MDRE could transport 1,200 tons of cargo from the LEO station to the SMF, exceeding the capability of today's upper stage rockets by a factor of over one hundred.

In NASA's scenario, the mass driver reaction engines are also used to transport raw lunar materials from the Moon to the SMF, and to retrieve asteroids, another potential source of raw materials in space. Elements that can be obtained from asteroids include iron, nickel, carbon, and hydrogen, all of which would be valuable raw materials for large scale space manufacturing. Astronomers have detected several near-Earth asteroids that contain such usable minerals and that could be accessible to a vehicle such as the mass driver reaction engine. Some of these asteroids travel in orbits so close to Earth that they can be reached with less expenditure of energy than is required to travel to the Moon. Despite these lower energy requirements, however, journeys to the asteroids would be several months to a few years in duration.

The NASA scenario calls for an ambitious asteroid

utilization scheme requiring MDRE's to depart for selected asteroids at six-month intervals. Each vehicle would support a minicolony of twenty-one persons and would require six months to reach its destination. After inserting the MDRE into the same orbit as the target asteroid, the crew would identify a one million ton fragment of the asteroid suitable for mining, and separate this fragment with chemical explosives. A space jeep would then be deployed to "de-spin" the asteroid fragment by winding it up with a cable and pulling in the opposite direction of the rock's rotation. After stabilizing the asteroid segment, the MDRE crew would maneuver it into an enormous bag made of a composite material resembling fiberglas. The MDRE would then begin its journey back to the Space Manufacturing Facility.

During the return trip, which could take up to three years, magnetic and heat separation would be used to extract desired substances from the asteroid fragment. About half the mass of the asteroid would be used as reaction mass fuel for the MDRE, leaving half a million tons of usable asteroid material upon the arrival of the MDRE at the Space Manufacturing Facility. From a typical asteroid, this would include about 100,000 tons of iron and nickel, 50,000 tons of water, and 20,000 tons of carbon compounds. The remainder of the asteroid fragment would consist of less valuable materials, but could be used as shielding mass. Eventually, millions of tons of such "slag" materials will be required to insulate permanent space habitats so they can protect their residents from cosmic radiation and micrometeoroids.

A More Ambitious Scenario

NASA's 1977 scenario for utilization of nonterrestrial resources is extremely ambitious, but it is based on making practical use of existing space hardware and is aimed at manufacturing specific, practical products. Despite our lack of progress in space since this study was conducted, an even more dramatic program for using space resources was recommended a decade later by an independent commission appointed by President Reagan to identify long range space goals for the United States. In its 1986 report, the National Commission on Space proposed that the U.S. develop capabilities to utilize lunar, asteroidal, and Martian resources by the third decade of the twenty-first century. The Martian moons Phobos and Deimos are cited by the space commission as potential sources of carbon, nitrogen, and hydrogen, all of which are relatively scarce on the Moon. Phobos is a particularly attractive target because it circles Mars at a distance of only 6,000 miles, and hence could serve as "an ideal refueling depot" for spacecraft travelling between the Earth-Moon system and Mars. The first step toward exploration and utilization of these Martian moons will be taken in 1988, when the Soviet Union launches an international robotic prospector mission to Phobos and Deimos.

Most of the key technologies required to establish a human presence on Mars have already been developed or could be demonstrated at NASA's LEO Space Station and an early lunar base. The National Commission on Space proposed following Space Station and lunar base development by immediately using Phobos and Deimos as stepping stones toward the establishment of a large settlement on the surface of Mars, which would begin operations between 2015 and 2030. A Mars settlement could be used to obtain resources from the Martian surface, but its main purpose would probably be to support scientific exploration. By studying the Martian environment, scientists at a Mars base could help determine whether the conditions that created life on Earth are unique to our planet or relatively common cosmic phenomena.

A major element of the commission's proposed scenario

for Mars exploration and utilization is a "cycling spaceship," also known as an "Earth-Mars Orbiting Station." The cycling spaceship, an interesting hybrid of a transportation system and a space station, could be permanently situated in an orbit that periodically passes very close to both Earth and Mars. Since the cycling spaceship would remain in a fixed orbit, it could be much larger and more hospitable to a human crew than small, lightweight vehicles designed for efficient transfer from one orbit to another. Personnel and supplies sent from Earth to Mars would begin their journeys from the LEO Space Station on OTV's, which would rendezvous with the cycling spaceship during its closest approach to Earth. Most of the crew's life support needs during the five to seven month trip to Mars would be provided by the cycling spaceship, which could employ rotating elements to provide artificial gravity. Upon reaching the vicinity of Mars, the OTV would be refueled and used to transfer the crew from the cycling spaceship to an outpost in Mars orbit or on one of the Martian moons. A Mars lander would then be used to transport the crew to the Martian surface.

Proposals for using Martian resources in an early space manufacturing system have emerged largely as a result of a broadening of the rationale for space development. Most participants in NASA's 1977 study supported the ultimate space manufacturing enterprise goal of building large space habitats capable of supporting millions of permanent residents. But the principal near term justification offered for development of the Space Manufacturing Facility and its support systems was the fabrication of solar power satellites. While interest in solar power satellites has dwindled since this work was performed, most people who have supported the concept of space manufacturing remain convinced that it is in humanity's interest to establish an industrial base in space. Now that the broader goal of expanding the human presence throughout the solar system has emerged as a more fundamental and immutable motivation for space industrialization, the establishment of Martian settlements has gained increased favor.

The primary obstacles to achievement of such goals are economic, rather than technological. The space manufacturing scenarios outlined by NASA in 1977 and the National Commission on Space in 1986 are both far more ambitious than the activities that can be undertaken over the next few decades if NASA is forced to continue doing business under its current budget constraints. If the United States, unilaterally or in cooperation with other nations, were to make a commitment to large scale space industrialization, the basic technologies required to make large scale space manufacturing a reality could be developed within a period of about one decade. The funding required to develop these technologies and to establish a space manufacturing system would represent a smaller share of the national product of the United States than was devoted to Project Apollo, but would still require a significant increase over NASA's budgets of the 1970s and 1980s. While we don't yet know of any product that will yield an immediate economic return on such an investment, we do know that space industrialization is the key to the inevitable expansion of the human presence beyond the confines of planet Earth.

9

The Colonization of Space

When one is initially asked to consider the possibility of living permanently in space, images of an exciting but limiting existence usually come to mind. Thoughts turn to small communities of specially trained astronauts, living together in cramped quarters reminiscent of flying submarines or, at best, large hotels designed to roam about the heavens. While this might turn out to be a fairly accurate description of life on board NASA's Space Station, it falls far short of capturing the essence of what life might actually be like in a large scale space settlement. Within the next few decades, modern technology will offer us the ability to transport many of our Earthly comforts into space. The report of the National Commission on Space states that "Living in space will be practical even though for long term good health, people and the food crops that support them require atmosphere, water, sunlight, protection from radiation, and probably some gravity." In fact, life in space may eventually be safer than living on Earth, and "space colonists" could some day enjoy an even broader range of intellectual and recreational pursuits than their modern day counterparts on our home planet.

The Modern Concept of Space Colonization

By many accounts, the modern concept of *space colonization,* or large scale migration of people into space, originated in a classroom of Princeton physics professor Gerard O'Neill in 1969. To kick off a weekly seminar set up for several of his advanced physics students, Dr. O'Neill asked his pupils to consider whether the surface of a planet is the ideal location to support an expanding technological civilization. After performing some initial calculations, Dr. O'Neill and his students concluded that a rotating pressure vessel several miles in diameter could be constructed in space, using ordinary materials. They also computed that the number of such habitats that could be built might provide more than a thousand times the land area of the Earth.

Contrary to the classical science fiction idea of colonizing distant moons or planets, O'Neill and his students determined that artificial pressure vessels — "inside-out planets" — could have several important advantages over planetary surfaces. With the exception of Earth, no moon or planet in our solar system provides the right combination of gravity, atmosphere, and temperature to support human life. A pressure vessel in orbit about the sun could be designed to meet all of these criteria, and would have additional advantages inherent in being in free space. Such a habitat would be bathed in a continuous flow of solar radiation, providing an inexhaustible supply of energy. Transportation between free-flying space colonies would be much easier than travelling between planetary surfaces, where tremendous amounts of energy are needed to escape the force of gravity.

Intrigued with these preliminary conclusions, O'Neill and his students then addressed the issue of whether space habitats could be made psychologically appealing as well as physiological-ly acceptable. It was immediately decided that space colonies would have to be as Earth-like as possible if they were to be attractive to large numbers of people. It was therefore determined that space habitats should contain trees, wildlife, large bodies of water, and other natural surroundings. A concept for using mirrors to direct natural sunlight into space habitats was developed, and it began to appear that not only were space colonies technologically feasible, but that they could be made sufficiently Earth-like to appeal to a broad spectrum of our planet's population.

Convinced that this new concept of space colonization could hold important benefits for mankind, O'Neill tried to encourage a broader public discussion on the subject. His early attempts to spread the word beyond his classroom, however, were unsuccessful. O'Neill's first article on space colonization was rejected by several magazines, and it was not until late 1972 that he delivered his first public lecture on the subject, at Hampshire College. But the student response at Hampshire was very enthusiastic, and O'Neill was subsequently invited to speak at several additional campuses, including many on the west coast. Finally, in early 1974, O'Neill's first article on space colonization was accepted, by the magazine *Physics Today.* A few months later, O'Neill organized a small conference at Princeton, which was attended by Walter Sullivan, a reporter for *The New York Times.* When Sullivan's article on space colonization appeared on the front page of the *Times,* a wave of publicity followed, and O'Neill and his concept of space colonization began to receive widespread attention.

O'Neill's efforts to popularize space colonization coincided with the publication of *The Limits to Growth,* the widely disseminated book forecasting the rapid demise of human civilization due to overcrowding, shortages of food and energy, and pollution. When *The Limits to Growth* and its gloomy predictions were published in 1972, O'Neill was given the final missing ingredient needed to sell his concept: a practical rationale for space colonization. In the spring of 1975, O'Neill's second conference on space colonization was held at Princeton, and that summer NASA appropriated $100,000 for the first of three consecutive summer studies on space settlements. The following year, O'Neill published *The High Frontier-Human Colonies in Space,* an eloquent rebuttal to *The Limits to Growth.* By 1977, when the last of NASA's three summer studies was concluded, O'Neill and his colleagues had succeeded in convincing a great many people that space colonization was, at least technologically, a legitimate option for humanity.

Since NASA's last dedicated study of space colonization was conducted in 1977, a multitude of smaller scale studies have been conducted by the space agency, universities, and aerospace firms on subjects relating to O'Neill's work. All of these studies have confirmed the technological feasibility of space colonization, and have collectively provided us with a database that will be drawn upon when humankind some day decides it is time to expand beyond Earth's horizons. Significantly, these studies have left unaltered the essential features of the visionary concepts outlined by O'Neill in *The High Frontier.* In this early work, O'Neill introduces the reader to *Island Three,* a cylindrical habitat twenty miles long and four miles in diameter that is assembled in space with raw materials mined from the Moon and the asteroids. It rotates slowly to provide its residents with artificial gravity, reflects sunlight into the habitat interior with mirrors, and has a pleasant Carribean climate. Its inhabitants live in private homes and terraced apartments interspersed with dense parkland and winding, trout-filled streams. They pedal to work on bicycle paths, enjoy fresh fruits and vegetables that are always in season, and are linked with their friends and relatives on Earth with the most advanced forms of audio-visual communication. The population of Island Three: *ten million.*

While Island Three sounds like the setting for a science fiction novel or a product of some advanced civilization of the distant future, many of the basic technologies required for its construction have already been developed. O'Neill projected that such a colony could be established by the latter half of the twenty-first century, based on experience that could be gained during the next fifty years by building earlier and less elaborate habitats: Island One (ten thousand inhabitants) and Island Two (one hundred thousand to one million residents). The technologies needed to assemble and populate settlements of this scale fall into four principal categories: space transportation, extraterrestrial mining and processing, construction of large space structures, and regenerative life support systems. In the decade since publication of *The High Frontier,* terrestrial research and Space Shuttle missions have made considerable progress in demonstrating the feasibility of applying all of these technologies to the colonization of space.

Space Transportation

The settlement of space by large numbers of people will require the development of at least three different types of advanced space transportation systems: heavy lift launch vehicles (HLLV's) to launch people and materials into low Earth orbit, high-thrust orbital transfer vehicles to rapidly transport people from LEO to more distant orbits, and extremely large cargo-carrying transfer vehicles for moving the input and output of space factories around the solar system. Experience with the Space Shuttle indicates that it may be two or three decades before such systems can be developed and proven to be operationally reliable.

If a commitment is made to large scale habitation of space, the first of these advanced transportation systems that will be needed will be heavy lift Earth-launched vehicles. In the mid-1970s it was expected that the Space Shuttle could be the "workhorse" for launching the basic elements of a space manufacturing enterprise into space, but this was at a time when NASA still harbored expectations of flying over sixty Shuttle missions per year. Since it now appears that the maximum Shuttle flight rate will be between ten and twenty missions per year, the need for a fleet of HLLV's will probably arise far sooner than originally expected by O'Neill and the early proponents of space industrialization. NASA and the Department of Defense, recognizing the potential value of HLLV's for a variety of space projects, have already spent millions of dollars studying dozens of different concepts. From these studies a consensus is emerging that the HLLV should be at least partially reusable, capable of delivering 120,000 to 180,000 pounds of cargo to low Earth orbit, and available by the middle or end of the 1990s. In early 1987, NASA and the Defense Department began their most active campaign to get the HLLV developed. If this effort is successful, such a vehicle could be flying as early as 1993.

One key issue that has yet to be resolved is whether an investment should be made in making the HLLV a manned system. To support space colonization, such a vehicle would eventually have to carry hundreds of people into orbit on a single flight. This requirement is not likely to have much impact on the current HLLV development effort, which is aimed toward satisfying much less ambitious mission requirements. Fortunately, a manned heavy lift vehicle may not be needed until the later stages of large scale space industrialization. The several hundred people needed in space to develop the space manufacturing infrastructure could probably be launched into orbit on modified versions of the Space Shuttle, which might be able to accommodate up to several dozen people in pressurized cargo bay cabins. In such a scenario, a manned HLLV would not be required until the residents of the first large space colony were ready to relocate to their new homes. Alternatively, large numbers of people could be ferried into orbit by the *Orient Express,* a very advanced single stage-to-orbit space plane that is currently in early stages of development.

While development of a man-rated HLLV can probably wait for several decades, advanced interorbital transportation systems will be required at the earliest phases of large scale space development. High thrust OTV's for ferrying people beyond low Earth orbit will be needed to emplace and support the initial lunar base, even if it is a relatively small scale facility. The initial space manufacturing facility will also depend on OTV's to meet the transportation needs of its crew. To meet these early needs, which may require transportation of as few as two people, and advanced requirements, which may call for transport of over two hundred, the development of interorbital personnel carriers will probably be evolutionary in nature. The OTV's used to support the early stages of space colonization may be similar to the vehicles being designed by NASA for routine delivery of communications satellites to geosynchronous orbit during the 1990s. By employing modular components, multiple stages, or subsequent design modifications, these OTV's could be scaled upward in size as requirements for crew transportation evolve.

Systems used for unmanned interorbital transportation of large quantities of material are likely to be developed differently. These vehicles will probably bear little resemblance to the Space Station-based OTV's of the late 1990s, and their development is likely to require the greatest time and cost of any of the transportation systems used to support the industrialization of space. To achieve maximum efficiency, such vehicles will probably utilize advanced forms of propulsion, such as the mass driver reaction engines envisioned by space colonization planners of the 1970s. Another possible means of propulsion is solar energy, which could be captured by large solar arrays and used to power new types of engines such as ion thrusters. An even simpler concept under consideration is the "solar sail," which would literally drift through space on the power of "solar wind," the constant stream of particles emanating from the sun. Vehicles such as these would require the production of little or no propellants, but would be suitable only for transportation of cargo, owing to the long periods of time required for them to reach their destinations. Since such cargo-carrying transfer vehicles will not be needed until requirements emerge for transportation of extremely large quantities of material, they will probably represent the last element of the space colonization transportation infrastructure to be developed.

Extraterrestrial Mining and Processing

Space transportation technologies may be consuming nearly all of NASA's available attention today, but the colonization of space will be dependent on advancements in several other areas. One of the most important of these is the technology of processing nonterrestrial materials. It is difficult to conceive of large scale space habitation without the processing in space of raw materials mined from the Moon, asteroids, Mars, or Martian moons. The estimated mass of Island One, Gerard O'Neill's first generation space settlement, is over three million tons, the equivalent of one hundred thousand full Space Shuttle loads. Even with very advanced Earth-launched transportation systems, this quantity of material would take decades to launch at a cost that would probably exceed half a trillion dollars. When the environmental impact of removing this much material from Earth is considered, in terms of both resource depletion and exhaust from cargo-carrying rockets, it becomes evident that space colonization would be a practical impossibility if Earth were the only source of raw materials for such an enterprise.

Use of extraterrestrial materials may in fact become desirable long before a settlement the size of Island One is established. An early indication of the value of space resource utilization was provided by the 1977-1980 study of solar power satellites conducted by NASA and the U.S. Department of Energy. This study was conducted with the groundrule that Earth would be the source of all materials required for the construction of these mammoth satellites. Under this assumption, it was concluded that construction of sixty such satellites could result in shortages on Earth of such materials as mercury, tungsten, arsenic, gallium, and silver. The near-continuous space launches required to get this material into orbit would inject large amounts of water and carbon dioxide into the Earth's upper atmosphere, depleting the ozone layer that protects life on our planet from harmful solar radiation. Due largely to the cost of these launches, the total price tag for establishing sixty power satellites was estimated to be nearly one trillion dollars. The real conclusion of the NASA/DOE study: if very large space systems are ever to be economically and environmentally acceptable, they will probably have to be constructed in space from nonterrestrial materials.

Over the past few years, NASA's long range planners have expressed renewed interest in extraterrestrial mining, a subject NASA has not studied in depth since the late 1970s. This resurgence of interest is due in part to the growing realization that Earth launch costs will remain prohibitive for quite some time. Even after space systems are permanently based in orbit at NASA's Space Station, they will remain expensive to operate if they continue to be dependent on Earth for logistics support and resupply. Since the demand for space based OTV propellants could be sufficient to warrant extraction of liquid oxygen from the Moon within the first decade of the next century, this particular application of extraterrestrial resource utilization has received the greatest amount of attention in recent studies. Liquid oxygen has already been extracted from Moon rocks brought to Earth by the Apollo astronauts, utilizing relatively simple processes that could be employed in an early lunar factory. A lunar base capable of extracting liquid oxygen from lunar ore and storing the propellants in large tanks could probably be established at a cost equal to or less than NASA's Space Station. Such a facility could be expanded as needs for additional extraterrestrial products emerge.

As lunar mining and processing operations evolve, it will eventually become attractive to process raw material in free space, rather than on the surface of the Moon. Factories in free orbit will be able to use solar energy continuously, a feat that cannot be accomplished on the Moon because of the two week lunar day-night cycle. Also, it will be easier to transport the products of orbiting factories to LEO and other orbits in the Earth-Moon system. For orbiting factories to economically process lunar materials, however, the percentages of lunar ore that are converted into usable products will have to be large, so that the energy expended to launch raw ore from the Moon is not wasted. Permanent space settlements, which will use the solid waste products of space industrial processes as shielding mass, may therefore be the first projects to justify movement of the initial factories into free space.

To obtain from the Moon the 3.6 million tons of material required to build a 10,000 person space settlement, half a million square meters of the lunar surface would have to be excavated. Even the full scale Space Manufacturing Facility designed during NASA's 1977 summer study would have to operate at full capacity for six years to process this quantity of ore. Space colonization may therefore require the retrieval and processing of asteroids as well as lunar mining. Use of asteroids for raw materials will probably require the emplacement of factories in free orbit, although the possibility of crashing asteroids into the Moon and processing them on the lunar surface has also been explored.

Once orbital space factories are established, it will probably become practical to keep them separated from the general living areas of permanent space settlements. Residents of space colonies who work in factories may have to commute through hundreds of miles of space to get to their jobs, although this distance would be traversed in a matter of minutes by advanced OTV's. Separation of industrial plants from living areas will enable space manufacturing to be carried out in the weightless condition of free space, rather than within the artificial gravity of space settlements. If located within a space settlement, a space factory might also create noise and pollution problems, and would take up valuable room that could be better used to create a more comfortable environment for the inhabitants of the settlement. Eventually, many of the goods manufactured by orbiting space factories might be delivered to Earth for use on our home planet. If space manufacturing ever becomes economically competitive with terrestrial manufacturing, it is conceivable that many polluting industries could eventually be moved into space, helping to protect Earth's biosphere.

Construction of Large Space Structures

Once sources of extraterrestrial building materials are established, the techniques used for assembling structures in space will differ dramatically from construction techniques utilized on Earth. Owing to the lack of gravitational stress, structures assembled in orbit can be designed to be much thinner and lighter in mass than structures we are accustomed to seeing on Earth. A grid of solar cells on an array in space, for example, might be as little as fifty millionths of a meter in thickness. NASA has already performed Space Shuttle experiments that have demonstrated that beams one hundred feet long and the thickness of aluminum foil can retain their rigidity in the space environment. If techniques for constructing ultra-thin space structures can be perfected, the costs of building and transporting such structures can be reduced significantly. One of the primary functions of NASA's Space Station will be to serve as our initial base for the assembly of large structures, which will be used for such applications as communications platforms, radio astronomy antennas, and solar arrays.

These early space construction activities will also help to develop the basic tools and techniques that will be needed for the building of permanent space settlements. A variety of manned and unmanned maneuvering vehicles, teleoperated robotic devices, and zero gravity hand tools will be in routine use by the turn of the century, and many of these assets will remain useful well into the era of space industrialization. Construction of habitats capable of housing thousands of people, however, will pose additional challenges, requiring advances far beyond those that will be achieved during the first few years of human residence in space. Unlike early space structures, permanent space habitats will not be designed for minimum weight. The primary design considerations for space settlements will be human safety and comfort.

The most demanding requirement for space settlement construction will be the protection of inhabitants from cosmic radiation. The Sun's cosmic radiation is composed of about 87% protons and 13% heavy nuclei, the latter of which have a high ionizing charge and pose a particularly grave threat to human life. To limit continuous radiation exposure among residents of a space settlement to levels within current U.S. safety standards — between 0.5 and 5 rem/year — about 280 grams of shielding material will have to be provided for every square centimeter of habitat surface area. For protection against occasional solar flares, which release still more radiation than the normal continuous level of cosmic radiation, an additional 220 grams per square centimeter of shielding mass will be required. To provide this level of radiation protection using the waste materials from space factories, a permanent space settlement will have to be surrounded by a layer of shielding at least six feet thick, which will comprise over 80% of the total structural mass of the habitat.

The need to maintain a suitable atmosphere within space settlements presents another major design constraint. Based on purely economic considerations, an oxygen-rich atmosphere for space settlements would be very desirable. It would permit space habitat atmospheric pressures to be maintained at lower levels than on Earth, reducing the structural mass required for containment of the atmosphere. It would also reduce the required quantities of inert gases such as nitrogen, which will probably be many times more expensive to obtain than lunar oxygen. With regard to safety, however, oxygen-rich atmospheres would be very undesirable because they would increase the danger posed by fires. To provide a safe environment, the atmosphere within space habitats will have to consist of 50% to 80% nitrogen. For an atmosphere of this composition to be breathable, it will have to be maintained at a pressure at least half as great as that of Earth, requiring a structural support mass of up to 100,000 tons for a 10,000 person habitat.

Another significant design consideration that entails a tradeoff between economy and safety is the artificial gravity level maintained within space habitats, which will dictate the rate at which the habitat structures must be rotated. If habitats are not rotated at a sufficient rate to produce Earth-normal or near Earth-normal gravity, then its residents could suffer such health problems as bone decalcification and muscle atrophy. Maintaining higher gravity levels, however, will increase structural mass and cost, and will be particularly expensive if a habitat's radiation shield is rotated along with its internal structure. In such cases, higher gravity levels could add several hundred thousand tons to the total mass of a 10,000 person habitat. To balance the need for safety and comfort with the desire to reduce construction costs, the gravity level within populated areas of space colonies will probably have to be kept between one half and three quarters of Earth's normal gravity level. In areas closer to the habitat's axis of rotation, the level of gravity will be much lower. Proponents of space colonization enjoy speculating about the new forms of recreation that might be enjoyed in such low-gravity regions by residents of space settlements.

Within the area of human factors engineering, the most obvious consideration affecting the construction of space habitats is the amount of personal space allocated to its residents. Astronauts during the Mercury, Gemini, and Apollo programs were able to work for several days at a time within extremely small space capsules, which offered as little as one cubic meter in volume per person. The Skylab astronauts, whose missions lasted two to three months, enjoyed about one hundred cubic meters per person, which may ironically be more than is initially offered on NASA's budget-constrained Space Station during the 1990s. For permanent residents of a large space settlement, it would be desirable to provide at least two thousand cubic meters of space per person, although hardy pioneers may be able to cope with about half as much living space during the early years of space colonization.

NASA's studies of these various design constraints have resulted in the conclusion that O'Neill's Island One space settlements, capable of housing 10,000 persons, will probably be spherical and about one mile in diameter. Intermediate sized habitats, with populations of 100,000 to one million, should be toroidal, resembling the rotating-wheel space station concepts popularized during the 1950s and 1960s. The largest space colonies, with populations of up to ten million, will probably be cylindrical, the most logical shape for habitats whose internal area could approach the land area of Switzerland. Depending on the population and environmental factors just described, the mass of these settlements could range from less than one million tons to over fifteen million tons. During NASA's 1977 summer study it was estimated that the cost of constructing such habitats, once the basic elements of a large space manufacturing system were in place, would be between $100,000 and $500,000 per person.

Regenerative Life Support Systems

Once we have established the means to build permanent space habitats large enough to house the population of a major city, it will become necessary to outfit these settlements with the means to support human, animal, and plant life in a safe and pleasant environment. To achieve this objective, advanced life support systems will be needed to provide space colonists with air, water, food, provisions for waste disposal, and a generally healthy and comfortable environment. The "open-loop" life support systems utilized on the Space Shuttle and planned for NASA's Space Station require oxygen, water, and food to be launched from Earth in their final, usable form for immediate use by astronauts. With these systems, waste products are dumped into space or packaged and returned to Earth. Permanent space settlements will require much more sophisticated regenerative life support systems, which will naturally transform most or all waste products into substances that are eventually used by some life form.

The average person living in a space colony will consume

about 450 pounds of food, 700 pounds of oxygen, and 1,400 pounds of drinking water each year. An additional 15,000 pounds of water will be needed by each inhabitant of a space colony annually for sanitary and household purposes. Based on these requirements, it would be a practical impossibility to support more than a few dozen individuals in space with resources delivered from Earth, at least with launch vehicle capabilities of the foreseeable future. Even if transportation systems were developed that could deliver such large quantities of material to orbit, the launch costs would probably be in the tens of millions of dollars per year for each inhabitant of a space settlement. And if launch costs (both economic and environmental) were affordable, reliance on Earth for resupply would be contrary to the basic philosophies of people settling the new frontier.

For these reasons permanent space settlements will almost certainly employ completely closed ecosystems from their inception. Every space colony will have to provide its own food, which will probably be produced in compartments that utilize modern forms of agriculture and livestock breeding. Agricultural areas will probably be separated from living spaces so crops can be raised in ideal growing conditions, which may require high temperatures, humidity, and abundant exposure to sunlight. Experiments in advanced crop growing methods have indicated that an environmentally controlled space-farm could produce ten times as much food per acre as farms on Earth. With yields this high, about forty-four square meters of growing space will be needed for each space settlement inhabitant. With an additional five square meters per person allocated for raising of livestock, the total land area required to produce food for a 10,000 person space colony will be about one hundred acres.

To recycle the water and air used by space colonists, the provision of which will be one of the most expensive elements of space habitat construction, will require significant advances in life support system technologies. Even domestic wash water, which is relatively easy to recycle, will require systems of tanks, ducts, filters, membranes, and chemical additives that would be very expensive to ship from Earth or fabricate in space. We know that it is theoretically possible to use oxidation processes to convert human wastes and trash into carbon dioxide, water, and ash, but to date such recycling systems have only been built and tested on a small demonstration scale. Oxygen can be recycled naturally from carbon dioxide through plant metabolism, but detailed plans for the provision of air cannot be established until the optimum atmospheric composition for space settlements is identified. Recognizing our lack of progress in these areas, the National Commission on Space identified development of "long-duration closed-ecosystems" as one of the seven technology advancement priorities most critical to the future of the U.S. Space Program.

The Quality of Life in a Space Settlement

Technologies in the four areas just described could probably be developed sufficiently to support space colonization within the first one or two decades of the twenty-first century. Gerard O'Neill predicted in the mid-1970s that an initial Island One space settlement could be built within 13 to 20 years if a commitment to colonize space were made. Since it will probably be several decades before such a commitment is rendered, residents of the first space colonies will in all likelihood benefit from technological advancements that are impossible to predict today. Through careful planning, life in space settlements can probably be made as safe and as comfortable as living on Earth. Once this goal has been met, continuing attention will have to be devoted to providing for the psychological needs of space colonists, so that people can live comfortably and happily in completely man-made worlds thousands of miles from our home planet.

The thick shielding required to protect the inhabitants of space settlements from cosmic radiation will reduce the danger of habitat penetration by solid objects such as debris from space activities and micrometeoroids. Probabilistic studies have indicated that a space settlement is likely to be hit by an object large enough to cause catastrophic damage only once every several thousand years. If such an object were headed for a habitat, it could probably be detected and destroyed, or the habitat moved out of the path of destruction. Objects too small to be detected might occasionally penetrate a habitat, but the loss of atmospheric pressure due to such an event would be slow, giving emergency crews time to repair any damage to the habitat shell.

By relying heavily on natural means of recycling waste materials, the chances of injury or death due to failure of space colony life support systems should be remote. The first space settlements may initially be dependent upon a limited number of food and water sources, but as the population in space expands, the number of alternative resupply routes should multiply rapidly. If a space settlement's agricultural areas or water regeneration system were severely damaged, food or water could be obtained temporarily from another colony. Since residents of space settlements will not have to deal with hurricanes, droughts, or other unpredictable terrestrial phenomena, life in a space colony may actually be more stable than life on many areas of our home planet.

With their physiological needs routinely met through a combination of modern technology and proven natural processes, residents of space settlements may find that their quality of life will depend largely on their ability to adapt psychologically to living in space. Residential areas within an early space settlement will probably have a population density of about 64,000 persons per square mile, roughly equivalent to the poulation density of Manhattan or Rome. While modern urban planning techniques can be used to make the best use of the limited living space within a space habitat, the feelng of overcrowding may be amplified by the space habitat's enclosure, isolation, and distance from Earth.

To help make life in a space settlement as enjoyable as possible, attempts will be made to bring the best features of Earth into space. As envisioned by Dr. O'Neill, the interior of a typical space habitat will probably be filled with trees, wildlife, streams, and other natural features to make it resemble pleasant places on Earth. To help offset their feeling of isolation, space colony residents may be able to keep in close contact with friends and relatives on Earth via modern forms of audio-visual communication. Proponents of space colonization point out that residents of space settlements might also have opportunities that people on Earth will not enjoy. A space settler, for example, may be able to take frequent journeys around the solar system in private or public space vehicles, visiting other colonies, resort settlements, or space hotels in scenic orbits. Eventually, colonies with varying climates and geographic features can be built, offering people the opportunity to live in space settlements that are ideally suited to their preferences. Whether these new opportunities and conveniences will offset the disadvantages of separation from Earth and of living in a sealed, man-made environment cannot be predicted. Will space colonies be enjoyable as places to live? This cannot be judged until the first large scale attempts are made to occupy the heavens on a permanent basis.

10

The Challenge of the Space Age

"Let's solve our problems here on Earth before we spend money on the Space Program." Opponents of space activities have voiced this sentiment since the early 1960s. Such expressions of concern should not be unexpected; space projects absorb billions of dollars in funding that could be used to meet society's needs in many other ways. The complexity and high visibility of America's space activities help to conceal many of their benefits and make NASA a convenient target for budget trimming. But people who initially oppose space expenditures are frequently surprised when offered a view of the Space Program from a different perspective. Educating the public so space investments can be evaluated effectively is vital if we are to achieve the greatest possible benefit from space exploration and utilization.

The Cost of Space Development in Perspective

There is no denying the fact that space activities are expensive. The $250 million cost of a single Space Shuttle mission could feed fifty thousand families of four for an entire year. Alternatively, this amount of money could be used to build high quality housing for five thousand of these families. But are space expenditures even remotely responsible for poverty in America? NASA receives about $8 billion in funding each year from the U.S. Government, while the U.S. Department of Health and Human Services (HHS) has an annual budget of nearly $400 billion. The agency with greatest responsibility for solving problems here on Earth therefore spends almost as much in one week as NASA spends in an entire year.

If HHS would go through NASA's entire budget in seven days, one must wonder how many additional human welfare problems this agency could solve if the U.S. dismantled its Space Program. If an additional $8 billion per year *could* buy a substantial improvement in social welfare, one would then have to ask if NASA's budget is the best source for this funding. A two percent reduction in the Defense budget, which rivals that of HHS, would cover NASA's expenses for an entire year. Viewed from a different context, NASA's funding represents less than one percent of the federal budget. In other words, a one percent increase in federal taxes, which would cost the typical taxpayer about one dollar per week, would create just as much revenue as the total abolition of our civilian space program.

Another factor that should be considered when evaluating space funding is the purpose of the expenditures. Most government appropriations go toward meeting society's near term needs, such as national defense, social security payments, and criminal justice. But a small percentage of the federal budget is used for investments that yield long range social benefits, such as education and medical research. The Space Program falls into this latter category, so its benefits are less obvious than those of government programs whose impact is more immediate. Transferring funds from space to social services might seem attractive in the near term, but would prove costly in the long run if downstream benefits were compromised.

The Benefits of Space Development in Perspective

In all probability, the U.S. Space Program generates more publicity per dollar spent than any other government activity. This is why many Americans find it surprising that NASA receives only about a one percent share of the federal outlay. Since news coverage of the Space Program focuses on NASA's glamorous and expensive accomplishments, many citizens are also unaware of the practical benefits created by our space activities. Opinions vary as to the total return generated by American investments in space, but the Space Program must certainly rank as one of the government's most productive endeavors.

The practical benefits provided by satellite communications and remote sensing of Earth's environment are the most obvious and quantifiable returns on our space investment. The billions of dollars in business and widespread conveniences created in these areas certainly go a long way toward justifying the nation's space activities. Space communications and remote sensing have also benefited Americans by adding an element of stability to the international military situation. Satellites are used by the armed forces for communications, monitoring of compliance with arms control treaties, and defensive early warning systems. All of these activities serve as important deterrents to armed conflict, one of the greatest but least recognized benefits of the Space Program.

The Space Program has stimulated the development of new technologies in many other areas as well, generating economic returns the totality of which are impossible to calculate. The inventions of teflon and velcro are commonly attributed to the Space Program, but these advances are relatively minor compared with others that have been brought about by our space activities. Far more significant has been NASA's impact on computer technologies; the Space Program probably accelerated the development of hand calculators and personal computers by several years. The impact of such "spin-off" benefits on America's economic growth and technological leadership has been tremendous. Economists have estimated that every dollar spent by NASA returns at least seven dollars to the national economy, and this may be a conservative calculation.

Even less obvious than technology spin-offs are the benefits the Space Program provides in areas of basic research and education. Space activities have undoubtedly stimulated scientific progress by attracting young people to pursue careers in mathematical and scientific disciplines. More generally, students of all ages and in all fields have come to regard the Space Program as one of the greatest examples of the wonders that can be accomplished with the human mind and spirit. The Challenger accident was especially tragic in that it was to be the first space mission dedicated to America's schoolchildren. The national response to this tragedy underscored the great power of the Space Program to energize and motivate America's youth. As society desperately seeks ways to keep young people out of trouble and in school, these positive effects of our space activities should be remembered.

Benefits Yet to Be Realized

When attempting to place the benefits of the Space Program in perspective, it is important to recognize that we are still in the very early years of the space age. Just as the full benefits of New World exploration were not realized for centuries, we are probably many generations away from enjoying the ultimate return on our investments in space. But we can already see signs that the revolution created by satellite communications will soon be followed by newer and more powerful waves of industrial development in space. The creation of life-saving drugs and other miracle substances in space is probably only a few years away, and the technologies needed for large scale space manufacturing are well within reach. The day when space industrialization enables us to remove all polluting industries from the Earth's surface and atmosphere may be within the lifetimes of some of our planet's present inhabitants.

And as impressive as these benefits may be, the greatest contributions of the Space Program may turn out to be in totally unchartered regions of human endeavor. Could space perhaps become an arena for widespread cooperation and development of mutual trust among the world's nations? International cooperation in the development of science and technology represents one of our greatest hopes for peaceful cohabitation on this planet. It is difficult to conceive of a better avenue for accomplishing this than space exploration and utilization. Since the technologies required for space development are strikingly similar to those employed by the modern military, expansion of space activities could help dispel the economic arguments for continuation of the arms race.

As greater numbers of people are given the opportunity to experience space flight first hand, the collective conciousness of humanity will expand and an enlightened perspective on our home planet may finally begin to take hold. Space will provide us with a vast laboratory to test new ways of living and working together, and the lessons we learn in space can be applied to improving life for everybody here on Earth. Of all the gifts that the residents of Island One could bequeath to Earth, the greatest of all might be their help in showing us how people sharing a precious and fragile homeland can live together without conflict.

The era in which we live presents humanity with three great challenges: to live in peace, to bring economic prosperity to all people, and to offer tomorrow's generations an exciting future of physical and spiritual growth. During its relatively brief existence, the Space Program has emerged as a central force in our quest to meet all of these challenges. By breeching the bonds of our home planet, we have taken the tentative early steps of a race destined to become an advanced interplanetary civilization. The impact of the embryonic space age on our lives, already great, will expand and intensify in the years to come, as our horizons become as limitless as the Universe itself.

Selected References

Introduction

1. G. Harry Stine, *The Third Industrial Revolution,* New York, NY: G.P. Putnam's Sons, 1975
2. Ken Kelly, "Playboy Interview: Arthur C. Clarke," *Playboy,* July 1986

Section I

1. Daniel J. Boorstin, *The Americans: The Democratic Experience,* New York, NY: Vintage Books, 1973
2. Bill Callahan and Joseph Thesker, "Scientist: Space Program 'is in a Dangerous Phase'," The Tribune, San Diego, January 28, 1986

Chapter 1

1. Peter L. Smolders, *Soviets in Space,* New York, NY: Taplinger Publishing Company, 1974
2. "The U.S. Returns to Space," *Business Week,* June 20, 1983
3. John M. Logsdon, "The Space Shuttle Program: A Policy Failure?," *Science,* May 30, 1986

Chapter 2

1. *Program Manager's Review No. 4,* Downey, CA: Rockwell International Space Division (System Integration-Space Shuttle Program), October 24, 1973
2. Barbara Stone, *Development, Implementation, and Implications of Space Shuttle Pricing Policies,* doctoral dissertation submitted to Clayton University, January 1983
3. Wayne Biddle, "Space Shuttle Won't Soon Carry its Own Financial Weight," *New York Times,* April 10, 1983
4. "New Pricing Policy," *Aviation Week & Space Technology* (Washington Roundup section), August 5, 1985
5. Eliot Marshall, "The Shuttle Record: Risks, Achievements," *Science,* February 14, 1986
6. Ed Magnuson, "Fixing NASA," *Time,* June 9, 1986
7. Ruth A. Lewis and John S. Lewis, "Getting Back on Track in Space," *Technology Review,* August/September 1986
8. *Space Business News,* December 15, 1986
9. Joseph J. Trento and Susan B. Trento, "Why Challenger was Doomed," *Los Angeles Times Magazine,* January 18, 1987

Chapter 3

1. Kenneth Gatland, consultant and chief author, *The Illustrated Encyclopedia of Space Technology,* New York, NY: Harmony Books, 1981
2. *Defense Daily,* March 3, 1983
3. *A Study of Space Station Needs, Attributes, and Architectural Options,* San Diego, CA: General Dynamics Convair Division, NASA contract NASW-3682, 1983

4. E.L. Tilton, III and E.B. Pritchard, "Overview of NASA Space Station Activities," *Journal of the British Interplanetary Society*, Volume 37, 1984
5. *Space Station Definition and Preliminary Design Request for Proposal*, Houston, TX: NASA Solicitation No. 9-BF-10-4-01P, September 15, 1984
6. "NASA Will Have to Stretch to Meet $8 Billion Station," *Defense Daily*, February 20, 1985
7. Craig Covault, "Space Station Redesigned for Larger Structural Area," *Aviation Week & Space Technology*, October 14, 1985
8. Ben Bova, "President's Message," *Space World*, January 1987
9. Maura J. Mackowski, "Safety on the Space Station," *Space World*, March 1987

Chapter 4

1. Robert E. Johnston, "The Integrated Team: Key to Implementing Design-to Cost," presented at 34th Annual Conference of the Society of Allied Weight Engineers, Inc., May 1975
2. *Agreement Between the National Aeronautics and Space Administration and McDonnell Douglas Astronautics Company for a Joint Endeavor in the Area of Materials Processing in Space*, January 1980
3. Michael C. Simon, "Space Station Design-to-Cost: A Massive Engineering Challenge," presented at the 44th Annual Conference of the Society of Allied Weight Engineers, Inc., May 1985

Chapter 5

1. Delbert D. Smith, *Communication Via Satellite-A Vision in Retrospect*, Boston MA: A.W. Sijthoff, 1976
2. *Geostationary Platform Systems Concept Definition Follow-On Study*, San Diego, CA: General Dynamics Convair Division NASA contract NAS8-33527, 1981
3. John Curley, "McDonnell Douglas Test Aboard Shuttle Reflects Move Toward Biotechnology Line," *Wall Street Journal*, August 29, 1983
4. Michael C. Simon, "Commercial Use of Space," presented at the Space Shuttle/ Space Station Business Opportunities conference sponsored by Pasha Publications, March 1984
5. Randall Rothenberg, "Space Hustlers," *Esquire*, May 1984
6. Daniel Osborne, "Business in Space," *The Atlantic Monthly*, May 1985
7. Peter Gwynne, "Rethinking Space Business," *High Technology*, June 1986

Chapter 6

1. Klaus P. Heiss, "A Private Initiative to Fund Orbiter V," presented to the Congressional Space Caucus, March 16, 1982
2. Craig Covault, "NASA Planning for Shift of Space Shuttle Marketing Operations," *Aviation Week & Space Technology*, November 1, 1982
3. *Civilian Space Stations and the U.S. Future in Space*, Washington, DC: U.S. Congress, Office of Technology Assessment, OTA-STI-241, 1984
4. David E. Sanger, "NASA Will Launch Earth Satellites Using its Rockets," *New York Times*, March 12, 1986
5. Michael C. Simon, "Utilization of Government Incentives to Promote Commercial Space Station Development," presented at the 37th Congress of the International Astronautical Federation, 1986

Section III

1. "Keyworth Sees No Justification for Space Station," *Defense Daily*, February 16, 1983

Chapter 7

1. Donella Meadows, *The Limits to Growth*, New York, NY: The New American Library, 1972
2. Thomas Heppenheimer, *Colonies in Space*, New York, NY: Warner Paperbacks, 1977
3. Sanford D. Mangold, *The Space Shuttle-A Historical View from the Air Force Perspective*, Maxwell Air Force Base, AL: Air Command and Staff College, Air University, report number 83-1540, April 1983
4. Bruce Murray, "Whither America in Space?," *Issues in Science and Technology*, Spring 1986
5. James A. Van Allen, "Myths and Realities of Space Flight," *Science*, May 30, 1986

Chapter 8

1. John Billingham, William Gilbreath, and Brian O'Leary, editors, *Space Resources and Space Settlements*, Washington, DC: NASA SP-428, 1979
2. J. Peter Vajk, Joseph H. Engel, and John A. Shettler, "Habitat and Logistic Support Requirements for the Initiation of a Space Manufacturing Enterprise," in NASA SP-428, see Chapter 8, reference 1
3. Brian O'Leary, Michael J. Gaffey, David J. Ross, and Robert Salkeld, "Retrieval of Asteroidal Materials," in NASA SP-428, see Chapter 8, reference 1
4. D. Bhogeswara Rao, U.V. Choudary, T.E. Erstfeld, R.J. Williams, and L.A. Chang, "Extraction Processes for the Production of Aluminum, Titanium, Iron, Magnesium, and Oxygen from Nonterrestrial Sources," in NASA SP-428, see Chapter 8, reference 1
5. Michael C. Simon, "A Parametric Analysis of Lunar Oxygen Production," *Lunar Bases and Space Activities of the 21st Century*, Houston, TX: Lunar and Planetary Institute, 1985
6. Michael C. Simon and Raymond J. Gorski, *Economic Implications of Space Resource Utilization Technologies*, San Diego, CA: Earth Space Operations under NASA contract T-9058K, 1985

Chapter 9

1. Stewart Brand, editor, *Space Colonies*, San Francisco, CA: Penguin Books, 1977
2. Richard D. Johnson and Charles Holbrow, editors, *Space Settlements-A Design Study*, Washington, DC: NASA SP-413, 1977
3. Edward Bock, Fred Lambrou, Jr., and Michael Simon, "Effect of Environmental Parameters on Habitat Structural Weight and Cost," in NASA SP-428, see Chapter 8, reference 1
4. Gerard K. O'Neill, *The High Frontier*, Garden City, NY: Anchor Press/Doubleday, 1982
5. National Commission on Space, *Pioneering the Space Frontier*, New York, NY: Bantam Books, 1986

Index